The Spymaster of Baghdad

The Spymaster of Baghdad

A True Story of Bravery, Family, and Patriotism in the Battle Against ISIS

Margaret Coker

HARPER LARGE PRINT
An Imprint of HarperCollins*Publishers*

All photographs in the book's insert are courtesy of the al-Sudani family, except the photograph of Abrar al-Kubaisi, which is courtesy of the Falcons Intelligence Unit.

FIRST HARPER LARGE PRINT EDITION

ISBN: 978-0-06-306320-4

Library of Congress Cataloging-in-Publication Data is available upon request.

21 22 23 24 25 LSC 10 9 8 7 6 5 4 3 2 1

To the Iraqis who work tenaciously
and bravely to improve their homeland.
May your sacrifices not be in vain.

To the editors whose red pens and
advice have made me a better writer.

To Craig, my huckleberry friend.
As two drifters off to see the world, you've been the
best companion this gal ever hoped for.

Contents

Author's Note

From ancient Mesopotamia to modern times, military commanders and political leaders have praised the stealth, patience, and guile of spy craft when used in defense of their homelands and to achieve military victories.

During the recent wars fought in Iraq against Al Qaeda and the Islamic State, this conventional wisdom has never been more apt. In an era when national armies have the most technologically advanced weaponry in the world, killing terrorists is easy. Finding them is often the biggest challenge. Despite this, most of the books written about the disastrous U.S. invasion of the country in 2003 and its aftermath have been told through the lens of military officers, soldiers, and policy makers, and through their checkered attempts to stabilize the nation after the overthrow of Saddam Hussein, reform the Iraqi political system, and fight the militants who terrorized the nation and its fragile new government. These tales, often gripping and powerful accounts of how individual army and marine units fought, died, or survived their deployments, end in a common theme: how the political and security chaos created by the U.S. invasion fueled the radical Islamist propaganda espoused by Al Qaeda's founder, Osama bin Laden, and the group's first leader in Iraq, Abu Musab al-Zarqawi, as well as his successors.

What most of these narratives leave out is the separate fight that raged in the shadows of the great military battles: the cloak-and-dagger work of spies to disrupt and dismantle the terror cells that were killing thousands of Iraqi civilians and American soldiers, and to capture the leaders who directed these evil acts. This omission is partially by design: many authors of books about the war on terror hail from military or policy backgrounds and understandably want to burnish their own reputations and histories. But this omission also has to do with the nature of the intelligence world itself, where the best, most effective counterintelligence work can only be accomplished out of the spotlight.

Few would describe Baghdad as glamorous as Casablanca was in the 1930s or Berlin during the

Cold War, but since 2003, in the aftermath of the U.S. invasion, it became, like those two cities, a magnet for spies. Intelligence agents from around the world descended on the ancient city ravaged by decades of Saddam's dictatorial misrule and security chaos, in part because of the rising international concern about the burgeoning Salafi jihadi menace posed by Al Qaeda, which by the mid-2000s had turned Iraq into its global terrorist headquarters.

Amid this intrigue, an unlikely man emerged among Iraq's new security agencies as a key player in identifying and infiltrating Al Qaeda's networks.

Abu Ali al-Basri had spent most of his adult life on the run from Saddam's secret police as part of the political opposition that had worked to bring down his dictatorial regime. Like most Iraqis, he had grown up reading the legends about his nation's glorious past as the cradle of civilizations. Ancient Arabs loved a thrilling spy tale, such as the legend of Gilgamesh, in which the heroic king kills his enemies thanks to ingenuity and espionage. Even the tales of the Prophet Muhammad describe how he sent undercover agents behind enemy lines to keep him and his followers safe from rival tribes. Abu Ali loved these stories of bravery and derring-do, but he never aspired to be a spy. His career as an intelligence professional began as a path to survival. During his years in the Iraqi underground, he honed an expertise in surveillance, cover stories, and dead drops, and especially for cultivating agents who might be in a position to relay lifesaving information. When al-Basri returned from a long exile to work for Iraq's first democratically elected prime minister after 2003, he had skills that could help counter the nation's newest national security threat.

Quietly, and controversially, al-Basri used his authority within the Iraqi government to stitch together an elite intelligence unit called al-Suquor, or the Falcons. He and his men worked independently from the newly reestablished security agencies that the Americans had refashioned for Iraq after 2003, institutions that cost billions of dollars in U.S. taxpayer funds but were failing in the fight against terrorism. The burgeoning spymaster worked, first, out of a makeshift office in a remote corner of the prime ministry complex in downtown Baghdad, and then, later, from a nondescript building along a pitted dirt road near Baghdad's International Airport. From there he would launch missions to hunt the Sunni Islamist militants, and then work to turn those captured by the Falcons into high-level informants. The technique, unlike other Iraqi intelligence services who relied on brutality and torture, developed high-level, actionable intelligence, making him and his unit one of the U.S. military's closest counterterrorism allies in the Middle East.

Not that anyone would know of the Iraqis' reputation from the U.S. Army's official history of the Iraq War—which covers the counterterrorism struggles from 2003 until the American forces withdrew in 2011—a time period during which the Al Qaeda threat exploded like a virulent plague across Iraq before being almost completely wiped out. The Falcons are absent from these annals, as they are from later newspaper reporting covering the period from 2011 to 2013, when Al Qaeda's leaders regrouped under the helm of Abu Bakr al-Baghdadi into a new formidable force called the Islamic State in Iraq.

When al-Baghdadi launched his blitzkrieg across southern Syria and northern Iraq in June 2014, massacring thousands of Iraqis and seizing control of more than four million inhabitants, few world leaders had expected this catastrophic rampage, or even knew the name of the man who had declared this war on the Western world. That's despite the multiple warnings that the Falcons' spy chief had sent up his chain of command and to his international partners.

Yet even after his American partners withdrew from Iraq, abandoning their round-the-clock electronic surveillance over Iraq's terrorist cells, Abu Ali al-Basri kept watch. He spent long days and nights in his unassuming, cramped office inside a converted five-room breeze-block building in the prime minister's compound in Baghdad's Green Zone, updating the files of the terrorist leaders who were still at large. Without the massive American network of telephone and internet data, the Iraqi spymaster had to rely on a growing network of human sources, both within the jihadi community and through extended family networks in Iraq. In the world of spies, HUMINT, or human intelligence, can yield as many rumors as high-grade intelligence. But in the early summer of 2014, one of these human agents told the Falcons that the Islamic State had set up training camps in the western Iraqi desert in advance of an ambitious operation to establish a religious state. Abu Ali had the war plan, but he didn't know the exact launch date for the Islamic State's military invasion.

When the Americans returned to Iraq as the lead partner in the international coalition working to defeat the Islamic State, the Falcons resumed their close counterterrorism partnership, but the Iraqis were also emboldened to act on their own.

From early 2003 to 2019 I reported from Iraq, chronicling long periods when Baghdad and its surrounding countryside was a kaleidoscope of horror. After years of sectarian fighting and terrorist bombs, the city had become a synonym for murder and mayhem. Unclaimed bodies stacked up in morgues, too disfigured to be identified; death squads roamed the streets; and terror attacks were so common that every day, when parents went to work, they could not be sure they would live long enough to return home in the evening and see their children again.

But the situation in Baghdad had never looked as dire as it did in the summer of 2014, after the Islamic State had seized one-third of Iraqi territory, decimated the armed forces, and advanced to a front line just fifty miles north of the capital. The city was in a panic, diplomats had initiated evacuation orders, and residents feared they would be left to a fate much like what occurred when the Mongols raped and pillaged their way through Baghdad in the thirteenth century.

I had known about Abu Ali al-Basri from reporting assignments in Iraq for the *Wall Street Journal,* both before and after the Islamic State blitzkrieg, but I had no inkling about his and the Falcons' exploits until 2017. That year, I was back in Baghdad working for the *New York Times* and was amazed at the city's transformation. In the north, the Iraqi army was still fighting the Islamic State and the terror group was continually threatening to launch waves of attacks inside the capital. But Baghdad was safer than it had been since the U.S. invasion. New cafes were opening weekly. Families strolled through riverside parks dotted with repaired playgrounds without the fear of a terrorist attack. Young men and women packed nightclubs to hear live rock music and flirt. How, I wanted to know, had the city avoided spiraling back to its bloody past when, a decade earlier, Al Qaeda had made Baghdad synonymous with murder and mayhem?

For months I asked dozens of Iraqi and American officials who had succeeded in making the Iraqi capital so safe, but no one could give me an answer. The one man who might have an answer—Abu Ali al-Basri, who by then had been named the head of counterterrorism for the national intelligence agency—had ignored my long-standing request for an interview. But to paraphrase Sun Tzu, the revered Chinese military strategist, each secret should not be revealed before its appropriate time.

Out of the blue, on one blustery day in March, Abu Ali invited me to his secluded offices on the western outskirts of the city. We sat in an antechamber of his main office and sized each other up while sipping several cups of sugary black tea. Abu Ali met many of my preconceived notions of a spy chief that had been formed by my long-standing predilection for John le Carré over Tom Clancy novels. He wore a smartly tailored gray suit and button-down shirt without a necktie, the type of anonymous fashion that legions of accountants or bureaucrats wear every day. His dark brown eyes were alert, but he displayed little emotion as he spoke quietly yet confidently about Iraq's security situation. Everything about his demeanor was self-contained; his hands remained in his lap or carefully clutching the tulip-shaped glass teacup, and he hesitated before answering my questions, cautiously choosing his words and giving them the full weight of his attention.

When the pleasantries were done, the spymaster got down to business. He had heard of my queries and wanted to set the record straight. "We have eyes inside," he told me, using the Arabic slang for a spy. "We have penetrated Islamic State."

That's when I first heard one of the most amazing tales of wartime espionage, in which, over the course of sixteen months, thirty suicide bombers were stopped and eighteen separate massive terror attacks on the Iraqi capital were foiled, each of which would have had the equivalent destructive capacity of the 1995 Oklahoma City bombing.

Over the following two years, I held more than two dozen meetings with al-Basri and members of his Falcons intelligence team. They told me about classified missions that add a rich and important layer to modern Iraqi history. They revealed the role they played in locating and killing the reclusive former leaders of Al Qaeda in Iraq, men who killed U.S. forces before their withdrawal in 2012 and who preceded al-Baghdadi; the network of informers who helped the unit track the rise of the Islamic State; their undercover operations, which allowed them a direct tap of information against their enemies during the massive ground and air war to defeat the terrorist organization; and harrowing accounts of planned terror acts against Baghdad, including a chemical weapons attack, which they successfully foiled.

Ultimately, my aim with this book is to recalibrate Iraq's history away from one that until now has centered on the Americans' sins, suffering, and victories, and to illuminate the admirable role that Iraqis have played and the sacrifices they have made on behalf of their country and the world in the war on terror.

A note on characters and names: the Arabic naming system does not always conform to the English system, with its first name, middle name, and family name. In polite company, guests generally address hosts and elders not by their first names, but by a common convention that conveys the value placed on parenthood and pride of place of a family's oldest child—

for example, Um Harith, which means the mother of Harith, and Abu Harith, the father of Harith. I use this convention in the book when it reflects the preferred form of address for several of my characters. For other characters, I use their preferred names that fit within the English convention of first and last names, rather than the longer Arab naming convention. English translations of Arabic names are notoriously inconsistent. In my book, I have used English spellings that my characters themselves prefer or that are the most widely accepted in Iraq.

Prologue

The skies glistened like dark onyx when, in late October 2019, the American Special Operations Forces team flew by helicopter into northwestern Syria to kill the most notorious terrorist in the world.

Once a marginal Islamic scholar from a midsize Iraqi town, Abu Bakr al-Baghdadi in the summer of 2014 became the scourge of the West when he led an army of religious zealots across northern Iraq and southern Syria, capturing territory equivalent to the size of the United Kingdom. He proclaimed himself caliph, the leader of the world's 1.8 billion Muslims, and for five years oversaw a reign of terror, enslaving tens of thousands of women, brutally executing his detractors, and inspiring terror attacks in countries as distant as Turkey, France, the United States, and

Sri Lanka. Al-Baghdadi's self-described Islamic State did something few had thought possible in the wake of 9/11: it eclipsed Al Qaeda in ambition, technical sophistication, and brutality. The movement boasted deep financial reserves, oil wells, and military research laboratories, and it had lured thousands of true believers to join the estimated six million Iraqis and Syrians trapped under its rule.

Most Iraqis fought back against this existential challenge to their nation. Hundreds of thousands of men, supported by the U.S.-led international coalition, volunteered to liberate their land in what became a grueling thirty-two-month ground war that included some of the most intense urban battles since the end of World War II. This alliance succeeded, albeit at a high cost, with an estimated ten thousand Iraqi security forces and at least twenty thousand civilians killed in the fighting.

In the shadows of the war, an elite American-Iraqi team had the task to hunt down and kill many of the top Sunni Muslim militant leaders, men who, like al-Baghdadi, had spent more than a decade after Saddam Hussein's ouster in 2003 fighting American troops in Iraq and the democratically elected Shiite-led government that succeeded him. But the self-anointed caliph had been an elusive prey. As his empire was collapsing in the fall of 2017, al-Baghdadi and his coterie of close relatives and trusted advisers fled the advancing Iraqi troops and slipped across the border into Syria, where the raging civil war enabled them to hide among those rebels who shared their extremist religious views and kinship ties.

On that crisp fall night two years later, the approximately five dozen U.S. commandos hoped that their long search was about to end.

When the Delta Force operators stepped from the helicopters onto the dusty, hard-packed ground, they were armed with some of the most advanced equipment in the world, including robots to defuse the type of deadly explosive booby traps for which the Islamic State had become notorious and cutting-edge forensic technology that could positively identify the man they had been ordered to kill. As the unit moved toward the remote farming compound, an additional secret weapon bolstered their confidence—insider information from one of al-Baghdadi's most trusted lieutenants.

The hunt for al-Baghdadi had started months earlier with the help of a little-known Iraqi intelligence unit called al-Suquor, or the Falcons. Earlier that summer, the head of the unit had word from a double agent who had a proven track record of giving reliable information. The asset told his spymaster the location of the

Syrian safe houses used by al-Baghdadi and his family. The tip led to an intense search whereby the Iraqi intelligence team tracked the terrorist leader through Syria, sending fresh leads and information to the Americans along the way.

When the U.S. commandos surrounded the farm where al-Baghdadi lived, they knew the details of its layout and the number of people normally inside with him, as well as the daily routines of the Islamic State leader himself.

The raid unfolded quickly. The American team called on the people inside to surrender peacefully. Four women and one man inside the building were killed when they failed to heed that command, while two men and at least eleven children were detained. Al-Baghdadi was not among them. The Islamic State leader had grabbed two of his children and dashed into an underground cellar. An American military sniffer dog chased him, and when the Iraqi was cornered, he detonated a suicide vest, killing himself and his children.

The explosion collapsed the room into which al-Baghdadi had fled, so the commandos dug through the broken slabs of concrete and choking sand and dust to recover pieces of his mangled body for proof that they had gotten their man. Fifteen minutes later, while the assault team gathered documents, computers, and phones from the compound, U.S. military technicians announced a positive identification from the human remains. "One hundred percent jackpot," the special operations team leader relayed over the radio.

Thousands of miles away, President Donald Trump and his national security team, who had been listening as the operation unfolded, were jubilant. The world's most-wanted terrorist was dead, a man who had sexually abused and tortured American humanitarian workers, who had offered religious justification for slavery, and who had inflicted unimaginable suffering on his fellow Iraqis. The American forces had recovered a ream of fresh evidence for intelligence officers to sift through for further clues about the Islamic State officials still at large. Almost immediately, the American president began hailing the death of al-Baghdadi as the most significant milestone in the war on terror.

But as the sun rose over the Iraqi capital the following morning, and the world's media were heralding the successful operation, Abu Ali al-Basri, a soft-spoken middle-aged Iraqi man who had spent decades perfecting the art of subterfuge and counterintelligence, was much more subdued. Sitting behind his expansive wooden desk that was covered by stacks of files, Iraq's counterterrorism chief was in a better position to assess the American raid. Not only was al-Baghdadi dead, but so, too, was Abu Ali's longtime asset, an Iraqi Sunni militant who had agreed to spy on the Islamic State leader in exchange for a promise from the influential Iraqi spymaster for protection for him and his family.

The news hit Abu Ali like a powerful uppercut to the jaw. It was the second time that one of his men had died in the line of duty, despite Abu Ali's vow to himself after the first shattering loss that he would never allow it to happen again.

Chapter 1
Blessings of an Oldest Child

Harith al-Sudani was born with large brown eyes, a broad forehead, and a weak chin, creating a sense of imbalance. Nothing about his looks or his upbringing in an eastern Baghdad slum would lead anyone to think he would become a hero. Nevertheless, his parents considered him a blessing—the answer to eight long years of prayers for an heir.

From his mother's point of view, Harith was a delight. He was an easy child who was always eager to please. He would fetch things for Um Harith while she was cooking and put his toys away, keeping the house tidy the way she liked it.

But as soon as Harith was walking and talking, his father, Abu Harith, began to worry that his son had little in the way of a backbone. It was unseemly for the oldest son of a father with ambitions like his to be such a mama's boy, motivated by a smile and a hug. In Abu Harith's view, tenderness was as useful as locusts at harvesttime. Moreover, in the Iraq of the early 1980s, it was dangerous.

Abu Harith, thin as a bean stalk and as gritty as the dirt in his father's wheat fields in southern Iraq, resolved early on to toughen up his son. A good father needed to teach obedience and perseverance, the qualities necessary for survival given that the al-Sudanis adhered to the wrong religious tradition, lived in the worst neighborhood, and had no political connections.

As Harith grew, his mother and aunts smothered him with kisses. They compared his light brown curls and sweet smile to those of an angel. But Abu Harith never once showed affection to the boy, not a single hug or pat on the head. When company came over, the patriarch sat in his lacquered wood armchair and lectured on the merits of tough love. A slap on the head if Harith spilled tea on the floor. A whack to his legs from a reed switch if Harith played too loudly. A whipping from the wooden broom kept next to the stove in the kitchen if Harith talked back. If his oldest son could endure what his father inflicted on him, Abu Harith reasoned, he could survive life in Iraq.

Abu Harith's zeal toward discipline wasn't unusual in Iraq. Saddam Hussein controlled the country, but the nation was full of petty dictators. In almost every family the patriarchs ruled with power born from Iraq's deep-seated tribal traditions that thrived on hierarchy and submission. As the eldest man of the extended al-Sudani family, Abu Harith assumed the birthright as patron, status that gave him power over a number of lives—his wife and ten children as well as the families of three of his younger brothers living in Saddam City. Each were honor bound to seek Abu Harith's approval on major decisions in life—whom they wanted to marry, where they wanted to work, and even what their children should study in school. In return, Abu Harith was obliged to arrange for those jobs and help pay for their weddings. And if any family member got into trouble with the police, Abu Harith would have to stand surety in the matter. In all this, Abu Harith harbored a single dream: that his eldest son would lift the al-Sudani family out of poverty.

Harith was, after all, good at schoolwork, so that was half the battle won. The crucial missing link was discipline and fortitude, traits that his father took it upon himself to teach. When the rest of his children misbehaved, Abu Harith rarely struck them. He transferred the punishments to Harith. With the broom in his hand, he would tell the boy what the eldest son of a family was for—to absorb the aches, pains, and worries of the others. Just like Abu Harith had always done.

This was the world that Harith was born into, the world he was meant to inherit, one where it was heroic just to survive a father's tough love, the interference of evil bureaucrats, and the whims of a dictator who saw people like the al-Sudanis as enemies of the state.

These are the blessings of the oldest child, Harith would always tell himself. Such as my blessings are.

To his wife, who blanched at her husband's treatment of their firstborn son, Abu Harith would say the beatings were for his own good. When the boy scored the highest marks in the entire district on the nationwide exam given to all high school seniors, a score that guaranteed him a seat at Baghdad's best university, Abu Harith felt vindicated. He walked around the neighborhood bragging that his son, the first member of the al-Sudani clan to reach college, would get a job in a well-regarded profession—like engineering.

As much as anyone in Iraq, Abu Harith knew that life didn't turn out exactly how one hoped for or planned.

In the mid-1970s, as the global oil boom turned Iraq into an international economic powerhouse, Abu Harith had dreamed of making a good life for himself and his young bride. The couple traveled from their agricultural village on the lower Tigris River one hundred miles north to Baghdad, drawn to the promise of the booming capital. Like several hundred thousand villagers, Abu Harith got off the train and found a cozy adobe home for rent in the budding new district on the eastern edge of the capital known then as Al Thawra, or Revolution City. The gleaming new streets and buildings were heralded as a step toward building a modern Iraq. But within a decade, Iraqis found themselves choking under the yoke of a new dictator, Saddam Hussein, and bleeding from a brutal war with Iran. The residents of Al Thawra, meanwhile, found themselves trapped in the undertow of Hussein's politics. The residents of the district were all Shiite, the branch of Islam practiced by a majority of the country but not their ruler. Saddam deemed his Shiite citizens potential fifth columnists, due to their shared religious identity with Iraq's mortal enemy, Iran.

Instead of becoming the vanguard of a new nation, Al Thawra residents were physically cut off from the rest of Baghdad by a fifty-foot-wide canal, a manifestation of the barrier that already existed between the lower-class Shiites new to the capital and the urbane families who had been in Baghdad for generations.

Families like the al-Sudanis could do nothing to reverse the slide of their political or social fortunes, except try to avoid politics and keep their heads down. Stuck in a ghetto that their new ruler had renamed Saddam City, they had neither the political connections nor the family wealth to move to another neighborhood. When a wealthy man from Jadriya, Baghdad's upper-class neighborhood, built himself a mansion along the banks of the Tigris River, his architect hired bricklayers from the denizens of Saddam City. When police looked for suspects in a robbery, they combed the streets of Saddam City. The only times when the government ignored the district was when the authorities were looking to recruit military officers, civil servants, or engineers. There was no quicker way to be disqualified for a job than the line on a man's identification card showing that his permanent residence was in Saddam City.

So it was that Harith and his brothers grew up without many heroes or inspirations for greatness. The country had only one Olympic medalist, and that was a 1960 weight lifter. The national soccer team had qualified for the World Cup once, but was knocked out in the first round. There was the singer Kathem al-Saher, who was the pride of Iraq, beloved by the dictator and the country alike. More often, however, the nation's one television station filled the airwaves with lurid tales of traitors and more traitors.

Like most of the other boys in their neighborhood, Harith went through his early life wary of what the future had in store. Each day along his route to school, he zigzagged through a web of alleyways, past his uncle's home, and across the empty hard-packed dirt lot where the neighborhood boys would play soccer, to pick up his two best friends, Ali and Wissam, so the threesome could walk to school together. Turning westward from Ali's home, the boys had to pass an abandoned two-story house that they, and everyone else in the neighborhood, were certain was haunted.

When Harith was still a toddler, the family who lived in that house disappeared one night. The father, mother, and three children were all gone. The following day, the neighbors all pretended not to have heard or seen anything, and soon afterward the entire neighborhood had erased them from memory. It was a time of war, when men were being drafted for the front lines, and Saddam believed that Iraq's majority Shiites would revolt against him on secret orders issued by the revolutionary government in Tehran. Jails were overflowing with people that Saddam's secret police had ripped from their homes and pulled out of Shiite mosques on suspicions based on paranoia rather than evidence. Few such prisoners were heard from again.

Iraqis had a fervent belief in the world of djinns, spirits that could be forces of good and evil. One winter afternoon, when Harith and his friends were thirteen, they were walking next to the abandoned building when Ali cried out. He swore that he saw an ifrit, a kind of ghost known to haunt ruins, walking around inside. No one else saw it. But no one doubted him, either. The family that had once lived there must have died in an unspeakable way, the boys reasoned, otherwise a relative would have come to claim the property or sell the land. Instead, month after month the colorless, crumbling structure slowly sank into itself, its broken windows falling out. None of the boys wanted to risk the wrath of the ifrit or be touched by the curse that had doomed the family. But they didn't want to admit to being scared, either. The next day, when Wissam suggested a different route home, the boys agreed with alacrity, without mentioning another word. But word of the ifrit quickly spread, and so, too, did the trio's refusal to walk on the street by the haunted house.

The neighborhood bully, a boy named Hussein who was a year older than Harith, saw an opportunity for mischief. Faggots, he yelled at the three friends. Look who is afraid of their own shadows.

Ali and Wissam obeyed the rules of the jungle. They decided not to antagonize the bigger boy. Harith, however, lost his cool. Fuck your mother, he shouted back. I'm no faggot and I'll prove it.

When school ended that day, Harith, Ali, and Wissam were joined by a group of at least ten other boys, including Hussein. In the five minutes it took to walk from the schoolyard to the haunted house, Hussein and his gang kept up a nonstop prattle of taunts, sure that Harith would break off and run away. Neither Ali nor Wissam remembered Harith saying a word. He was in a different zone. When they reached the abandoned house, Harith didn't even flinch—he walked swiftly to the sagging doorway. Standing on the mildewed floorboards, he peered inside, hesitated for a moment, and then walked in, disappearing from his friends' line of sight.

Minutes passed. Yet Harith had not reappeared. Ali's heart raced, faster than a rabbit's caught in a trap. Harith's foolhardiness was going to be the death of him. The djinns, he thought, had trapped his friend inside. Ali yelled for Harith. Wissam did as well. But inside the house was silence. You killed him, Ali shouted at Hussein. The djinns have taken him!

Wissam urged Ali to go find his father, who worked in a shop just a couple of streets away. Someone had to go into the house to find Harith, but neither one could drum up the courage to do it himself. Ali broke away and, just as he was reaching the corner, he heard a loud laugh. He turned to see Harith back on the street near Wissam.

Ya Ali, you donkey. Come back. I'm not dead. At least not yet.

Ali never did ask Harith what had sparked his act of bravery that day. Years later, as an adult, Harith told his friend that he had an inexplicable urge to see what he was made of. A journey into the lion's den, is what he called it.

On the rough-and-tumble streets of Saddam City, where corporal punishment was the rule, families had an incongruous obsession with poetry—a nostalgia that evoked the golden era of Islam, when Baghdad was the center of the world and promoted scholars, scientists, and artists.

Each Friday in the al-Sudani family majlis, the room in every Iraqi house reserved for guests, men would sit on the floor cushions spread across the room, sip tea, and listen to Abu Harith recite poems that he had memorized as a boy or read in the newspapers that week. Harith enjoyed these afternoons more than anything else in his week, enchanted by the rhythms of a couplet and the allegorical weave of sonnets. He'd dissect these readings like a mathematician, finding beauty in the form and meter as much as in the emotions the poems could elicit.

When he reached the sixth grade, Harith also found a way to profit from his hobby.

The all-boys middle school, Al Joulani, was a ten-minute walk from the al-Sudani home in Saddam City. It was housed in a nondescript low-slung concrete stucco building, one of thousands built under the central government's education drive in the 1970s and 1980s. Like most of the district's buildings, the school appeared old from the moment it was first opened. Baghdad's intense summer sunshine had bleached the canary yellow walls the color of egg yolk. Patched plaster in the hallways couldn't completely hide cracks along the ceiling joints caused by the humidity. In Harith's neighborhood, these flaws didn't matter much anyway. No school inspector would be coming to check the work of the contractors. The residents themselves weren't going to take it up with the authorities. Nothing good would come of someone raising a complaint, exposing themselves as potential troublemakers to those in power.

Saddam City's reputation made recruiting teachers difficult. So no one questioned why the boys at Al Joulani had an empty hour in their school day, unchaperoned except for the obligatory portrait of

Saddam Hussein peering down at them with his thick mustache and dead eyes. Without adult supervision, most of the boys went wild. Some organized wrestling matches. Others played games with string and spitballs. Many sat in groups discussing girls, who they might fall in love with, who might let them steal a kiss, and who might let them go further than that.

Harith, however, spent the hour alone at the narrow wooden desk that he normally shared with two other boys. At thirteen, Harith had established himself as the smartest boy in the class, a status attributable less to his natural cleverness and more to his father's discipline over homework. He was also good at writing poetry.

Abu Harith's control over Harith extended into all parts of life, dictating everything from what color pants he could wear—only brown and never black—to how many hours the boy could sleep. It was lucky for Harith that his father also considered poetry one of the foundations of a respectable education.

Harith sold the poems he wrote at school to classmates wanting to impress their sister's friend or a neighbor's daughter. Soon, word spread around the neighborhood that Harith's work attracted girls like bees to flowers, and his reputation was made. But for all the success of his friends' romances, Harith himself never had any luck wooing a girl. Ali and Wissam used to joke that the djinns from the abandoned house had cursed him when he was inside. Some days, Harith thought they might be right.

By the time Harith turned fifteen, he had no more time for sonnets. It was the mid-1990s and his father's small printing business—where Harith worked both before and after school—had gone bankrupt. After the disastrous war with the Americans, international sanctions tore apart the Iraqi economy. Meanwhile, the al-Sudani family had multiplied and someone had to help Abu Harith feed all ten of his children. So he arranged for his oldest son to work alongside his cousin at one of Baghdad's open-air wholesale markets.

Six days a week, the two young men would rise before dawn. Harith would put on one of his two pairs of brown pants, bought, like all the al-Sudani children's clothes, at the secondhand markets in Baghdad al-Jdeideh, a neighborhood south of Saddam City. He would button up the shirt that his mother had ironed for him the night before, take a piece of bread and white cheese for his breakfast, and eat it while he walked three miles to the Jamila market. For six hours, Harith and his cousin would haul seventy-five-pound bags of rice, flour, and sugar between delivery trucks and market stalls. Harith, who was short and stocky like his uncle, complained that his muscles felt like a tightly wound oud string ready to snap.

Because he was too poor to own a watch, Harith was spared the torture of counting down the minutes until the noon call to prayer, when the market closed. Before the stall owners would head to the mosque, Harith would take his daily wages from them and buy food to take home to his family, drums of vegetable oil, and burlap bags of dried chickpeas and bulghur. He would hitch a ride back to Saddam City, and, after dropping off his bags, he would walk to school.

After a month of this backbreaking labor, Harith's cousin surprised him. He had saved enough money for a pushcart with which the two teens could double the amount of cargo they were able to haul and thus double their wages. Their initiative became fodder for jokes among the traders. Harith became known as a guy who could sell dates to a date farmer, or a rug to a carpet merchant. Soon, the two had enough cash to go into the arbitrage business, buying bulk goods at wholesale prices and reselling them to retailers around Saddam City. Without really planning it, Harith was making a relative fortune. He was bringing home around $15 a day, a much better salary than his father ever had.

Early one morning, while he was getting dressed for work, he heard his mother tell her visiting sister how proud she was of him. Traders sought him out because he had a knack for deals. At home the family relied on him for clothing and food. As he crossed the bridge that separated Saddam City from Baghdad that morning, Harith walked taller in the wake of the rare approbation. On the way home, he didn't entertain his usual complaints of aching feet. But his mood soured when he saw his father in the courtyard, broomstick in hand. Somehow he had heard that Harith had done poorly on his last school exam.

You are nothing but a glorified donkey, Abu Harith told his son. Unless you graduate from university you will never be better than a beast. The words stung more than the beating.

On a bitter winter morning in February 1999, Harith walked to school for his university entrance exams. Heavy rains had flooded the streets and his shoes were soon coated with sucking, thick mud, their weight a constant reminder of the burden of his father's expectations. He walked the long way, bypassing the road on which the haunted house was located. He didn't want to tempt fate and attract bad luck.

Later that spring, when exam scores were published, Harith received the best marks in all Saddam City, results that guaranteed him acceptance to college. That day, Abu Harith received a nonstop flow of guests congratulating him on his son's achievement. The idea of praising Harith or giving him a gift for his success never crossed Abu Harith's mind.

In 2002, Iraq found itself wedged between the tectonic plates of history. The Americans were again determined to launch a war, and Saddam Hussein was not fit to prevent the catastrophe. The al-Sudani family was barely aware of the geopolitics consuming much of the world—they were waging their own epic battle.

The showdown had been building since Harith started attending classes at Baghdad University. Each morning, as always, he woke up at sunrise. He washed his face with a trickle of tepid water from the family's single bathroom, just off the kitchen. He ripped through the hot flatbread that his mother had placed on the tin serving tray, drank two small glasses of sweet, deeply brewed tea, and then boarded the minibus for the twenty-minute trip downtown, past the leafy boulevards of Karada to the green expanses of the Baghdad University.

Through the soaring concrete campus gates, Harith felt he had walked a flower-lined portal into another land, a place where people looked and spoke like him but lived a substantially better life. Sewage didn't run through their streets. Electricity didn't flicker off in their classrooms. Sidewalks were free of flat tires and rusty construction waste. Students expected life to offer them more than beatings, unlike in Saddam City. For the first time in his life, Harith had access to knowledge, music, alcohol, and girls. But not necessarily in that order.

The al-Sudanis weren't the type of family to recite religious verses, but Abu Harith had a fierce moral streak. In his mind, alcohol was bad because it revealed a person to be unreliable, undependable, and weak, character flaws that would stain the whole family. Dating was the same. If teenage boys and girls spent too much time together unchaperoned, djinns would tag along. They would become wild, unpredictable, and deceitful. Order would be upended and the family's reputation ruined.

When Harith walked into his first class at university, Introduction to Engineering, and slipped into the tight, wooden desk, he wasn't prepared for the shock. Unlike school in Saddam City, the university was coed. Harith hadn't been in such close physical proximity to women his own age in years. He couldn't concentrate on anything the professor said. His nose was filled with flowery perfume wafting from the young woman sitting directly in front of him. She smelled of orange blossoms and laughed like an angel, her chestnut-colored hair shaking and catching the light as she tossed her head back.

Harith didn't know her name, but he knew he was

in love. And he knew exactly how he would show it. When his classes were finished for the day, he went to Mutanabbi Street, Baghdad's famous street of booksellers, to buy thick, creamy paper and a calligraphy pen. He went home and wrote three masterpieces of intricately woven prose intending to woo the woman who had captivated his heart.

For three weeks, he came early to class and dropped a sealed letter on her desk. When she arrived, he watched surreptitiously while she picked up the missive and read his poems. Clearly curious, she would scan the room trying to determine who had left the notes. She never looked at Harith. On the fourth week, his final letter contained an ode by Nizar Qabbani, the famous modern Arab poet.

My Lover asks me:
What is the difference between me and the sky?
The difference, my love,
Is when you laugh
I forget about the sky.

At the bottom of the page Harith wrote a request to meet the following afternoon at the Abu Nuwas park along the banks of the Tigris River. It was time, he wrote, to introduce himself.

That night, anxiety kept him awake. Um Harith kept asking him if he was sick, mistaking his agitation. Harith felt like his every nerve was on fire, but he told her that everything was just fine.

Harith knew he was crossing several taboos by pursuing the woman from his class. During the four weeks of correspondence, Harith had cautiously asked about his classmate, Nisreen. Turns out, she wasn't Shiite. She wasn't even Arab. She was from a Kurdish family in Sulaymaniyah, a combination that, as far as his father was concerned, would make her as unsuitable a choice for his son as a chimpanzee.

For once, Harith ignored the nagging voice in his head, the voice of his father that he had always routinely obeyed. He told himself that if Nisreen came to their rendezvous, then she must be his destiny. Something his father couldn't interfere with, even if he tried.

The next afternoon, when Harith arrived at the park, Nisreen was already there. Her eyes widened when she saw him. Her parents had always told her that the residents of Saddam City were either radicals or animals. She had never suspected that poets lived there, too.

Harith had been afraid of being tongue-tied, but instead the words flowed out of him. He bought them pomegranate juice from a wooden stall near the July 14

Bridge. They talked about books and school and their families.

They promised to meet again, and again after that. Soon, his appointments with Nisreen took up more time than his schoolwork. His father complained that Harith was never home anymore, but the young man made up excuses about needing to study at the library in the evening, a good lie as it was not entirely untrue. In their home with ten rowdy children, he never had enough space or quiet to do that.

As fall moved to winter, Harith became drunk on this feeling of freedom. His hours with Nisreen made him believe that his life might take a different direction, one in which he didn't have to live in Saddam City, where he could choose his own wife and live happily ever after.

By springtime, his lack of interest in coursework bore consequences. When the Engineering Department posted the end-of-year-exam results, Harith's name was listed below the thick red line indicating who passed and who failed.

Suddenly, he was a man with a noose around his neck and the trapdoor had just opened underneath him.

No matter how violent life was in their neighborhood, the al-Sudani kitchen was always warm and full of loving energy. Before dawn, Um Harith would be up baking flatbread and brewing tea. She peeled vegetables all morning and carted jugs of drinking water from the neighborhood well. By the time Harith was in university, she had ten other children at home to feed and few tools to aid her.

The afternoon that Harith learned he had flunked his exams, he came straight home, eager to be surrounded by the comforting rituals of cooking that hadn't changed since he was little.

On Um Harith's two-burner range was a wide-mouthed steel cauldron simmering with chickpeas. Next to it was a tureen filled with white beans cooked tender in tomato sauce and a bit of goat meat for flavor.

Harith's sisters prepared the family room by spreading a plastic tablecloth decorated with red cherries on the floor. They poured laban, a thin yogurt drink, into plastic cups for the children and placed piles of fresh bread and washed herbs around the sitting area.

Harith wanted to tell his mother everything that had happened to him. How he had lost his heart and now his place at university. But he couldn't get the words out of his mouth. Um Harith was distracted with her multiple tasks, calling the children to come sit down and carrying the platters of food to the sofra, the cloth laid on the living room floor around which the family sat to eat their meals. She didn't notice anything wrong.

When his father came in and washed his hands, preparing for the meal, the television blared news in the background about America's plans to bomb the country. Harith's siblings sat around the edges of the sofra, each scooping food into their mouths with the fresh flatbread, laughing about a cartoon they had watched that morning.

Harith sat on his knees, twisting a tissue in his hands. He knew his task of delivering the bad news would be no easier for delaying it.

Father, I have some bad news, he said, the words rushing like a spring flood. I have flunked my exams. The college is kicking me out.

Um Harith waved a dishcloth in front of her face. She felt faint, as Harith continued to speak.

He told his parents that the reason for his poor academic performance was because he had fallen in love with a girl, a Sunni from Iraq's Kurdish region.

Abu Harith almost choked on his food. The television kept droning on, about American perfidy and doom. But the rest of the room was silent. They all knew that Harith's confession had done more damage than if an American missile had landed on their house.

Abu Harith screamed the same curses against his son that he had since Harith was little. But inside, his stoicism crumbled. He didn't sleep at all that night. Something had to be done to save his son's future and the family's reputation.

The next morning, he drove directly to Al-Jadriyah, to the university campus; his distraught demeanor helped coax the guards to allow him entry. Abu Harith sat in the corridor of the Engineering Department in abject protestation. He pleaded with the dean to reinstate his son. When he refused, Abu Harith tried to move higher up the administrative ladder, seeking out the university president himself.

It didn't matter that Abu Harith wore a gleaming dishdasha, the traditional ankle-length loose shirt favored by Arab tribesmen, and a crisply starched headdress. His deeply creased face and wiry frame lacked authority in the venerable buildings of Baghdad University. He sat for six hours on a wooden bench outside the office of the president's secretary. He watched a river of people stream past him, in and out of the important man's workspace. No one paid him any attention. His desperation hung around him like vultures over a dying beast.

Finally his mortification was too much. On the long bus ride home, Abu Harith's sour mood curdled and hardened into a deep bitterness. He couldn't see beyond the red blanket of shame that was settling on his shoulders and covering his family. When he walked through the front door, Harith was sitting in the corner, nibbling cookies that his mother had made for him.

The mood at the al-Sudani home was raw and sharp, and years later the family never tired of sucking on the hollow bones of the argument that ensued.

Abu Harith threw his shoes at his son, his derision unbridled. I always said you were no better than a donkey, and look at you now, he yelled. Eating food, just like a beast. Waving your dick at school, just like a beast. Every breath you take disrespects me. I wish you had never been born.

Harith had spent twenty years meekly accepting his beatings, but that evening, he snapped. Harith jumped up and raised his own hand against his father.

Son of a whore. For once, would you just shut up?

Um Harith screamed and collapsed in a dead faint. Her other children rushed into the room, alarmed at the outburst. Harith froze. Instead of striking his father, he moved toward his prone mother, rolled her onto the cushions and checked to see if she was breathing. Several al-Sudani siblings burst into tears, wailing that their mother was dead.

Harith felt a new surge of anger at his father, this time for putting Um Harith's health at risk. Your tyranny will be the death of us, Harith told his father. As long as we die silently, with our honor intact, you won't care. From this day on, I refuse to listen to your orders. I will marry Nisreen, no matter what you say.

Abu Harith's face turned purple. Was his son possessed by a djinn? He wasn't a superstitious person, not like his wife, but at that moment he believed what she had told him the night they had learned Harith's news. The Kurdish girl must have cast a spell on their boy. If he had been in his right mind, none of this would have happened.

If you marry that witch you will be the cause of your mother's death, not me, Abu Harith shouted. If you do this, I will disown you. I swear by my father's honor. You will be dead to all of us.

Harith's adrenaline evaporated as suddenly as it had appeared. His sisters were sobbing, begging their mother to wake up. Harith knew, like his father did, that he could not survive without the family. He would have no job, no place to live, and nowhere else to turn. His father would ensure that the rest of the al-Sudanis would cut him off too. A new terrifying thought emerged as well. Nisreen might not even agree to be with him, a penniless young man from Saddam City. He hadn't seen her since learning he had failed his exams. What was it that her parents thought of him and his neighbors? That they were animals, like his father had just described him.

Harith stormed out of the house to the small front courtyard, desperate to find some space to contemplate the corner his father had trapped him in. Abu Harith had the upper hand in this match of wills—but his son didn't want to admit it just yet.

A moment later, the front door opened. Harith's younger brother Munaf walked out and stood next to him.

Mom is awake. She's drinking tea, he told Harith. Come back inside now. She's calling for you. Don't disappoint her. Munaf led Harith back into the family room, patting him on the back.

Harith took one look at their mother, pale and helpless, propped up on pillows on the floor. He knew that if he defied his father, he might as well kill his mother with his own hands. Either way, he would be responsible for her suffering.

May God forgive me, Harith said softly. Father, you win. I will forget Nisreen.

Years later, when discussing the event, the al-Sudani brothers remember the moment in which something vital inside Harith died.

Chapter 2
A Chance for Freedom

The news spread like joy at a wedding, its potency increasing as it swept across the trash-strewn lots of Saddam City. Throughout Baghdad, every military barracks, every office of Saddam's secret police and ruling party had been abandoned. The torturers had fled, the dictator was in hiding, and American tanks were in the streets.

Harith's younger brother Munaf, now eighteen years old, couldn't contain his glee as he marched to downtown Baghdad with hundreds of his neighbors and friends from Saddam City in the warm April sunshine, joining the crowds cheering for freedom and democracy. The dictator's hulking statue in Firdos Square, in which he raised his right arm like a twenty-foot-tall curse, was torn down.

For the previous two weeks, while the U.S. military rained bombs on Baghdad and American soldiers advanced from Iraq's Shiite-dominated south toward the capital, the al-Sudani family and almost all of Saddam City's two million residents had barricaded themselves at home. They were reticent about the invasion. No one wanted to be fooled like they had been in 1991 when President George H. W. Bush had told the Iraqi people to rise up against their dictator, and they had, only to be abandoned by the Americans and slaughtered by the thousands. The betrayal was still fresh in 2003—few Iraqis were naïve enough to hope that the younger Bush would change their lives for the better.

But that clear, sunny morning, with birds singing in the date palm trees, Munaf reveled in the deep waters of optimism. After all, the impossible already appeared to have come true. Earlier in the day, across the canal separating Saddam City from Baghdad, he saw American tanks patrolling the city's east-west highway. It wasn't just a statue that had fallen. The regime was gone as well.

At lunchtime, Munaf had grabbed Harith and together they joined the flood of men, both young and old, gathering in the streets. The crowd had no leader, but everyone felt the same magnetic pull. The men of

Saddam City strode down Revolution Street, crossed the bridge spanning the canal, and streamed toward the city center, the heart of their ancient capital.

Along the way, shopkeepers who normally would give men like Munaf the cold shoulder handed out sweets as they walked by. Women waved at them from the upper floors of apartment buildings, trilling in celebration. Thousands of people from all across town were in the streets, churning and mixing in ecstatic groups, giddy with the unfamiliar euphoria of freedom. An elderly man wearing the flowing tribal robes of a Sunni sheikh kissed Munaf on the cheek, like he was his own son.

Ever since he was little, Munaf had harbored a secret dream about what he wanted to do with his life. Yet for just as long, Munaf had been taught that people of his station couldn't achieve their dreams. They couldn't breach the upper-class neighborhoods of Baghdad from their ghetto, to walk downtown like he had today, dressed in his favorite black T-shirt that showed off his chiseled physique. But today, it felt like working-class Shiites like him owned the city. Like no one could stop them.

Standing in the crowds of Firdos Square, feeling the tide of history crash upon him, Munaf thought that he might be able to live the life of his dreams after all, rather than the life dictated as much by his father as by the regime that had been swept away.

Later, over tea with Harith, as the city still celebrated, Munaf told his brother his secret wish. He wanted to help bring justice to Iraq, serve the people trampled by Saddam's brutality. He wanted to be a policeman.

Under Saddam's twenty-five-year rule, Iraqis had come to refer to their nation as the Republic of Fear due to the web of police agencies that had the sole purpose of protecting the regime instead of defending the citizenry against its abuses. Shiites were excluded from almost all jobs in the security forces because the Sunni regime believed they were a national security threat. No one in the al-Sudani family, or their social circles, had ever dreamed of being an officer. Some uncles and cousins had been drafted to fight on the front lines during the Iran-Iraq War, but they had been mere foot soldiers, cannon fodder, not leaders.

Even on a day full of hope, Munaf's dream was fantastical. No one in the ever-growing family could spare an ounce of care or effort for the desires of the third-born son. For his entire life, Munaf had been a pale afterthought in their father's fixation on bringing glory and prosperity to the al-Sudanis. For proof, he had only to look at the collection of photos hanging on the beige-colored walls of the room where the family gathered each day for meals. The largest portrait was of Abu Harith and his wife dressed in stylish Western clothes. Um Harith had a warm smile, her peaches-and-cream complexion glowing and her bouffant bob as stiff as starch as she held a light-haired toddler with curls and wide brown eyes—young Harith, gazing slightly beyond the photographer, as if looking to the future. Next to that photo, in an elaborate gold frame, was another portrait of Abu Harith, holding his somber firstborn as an infant. A third photo showed Mrs. al-Sudani with a leopard-spotted headscarf tied loosely over her dark brown hair while holding Harith with one hand and another infant, her second son, Muthana, on her lap. Other family memories immortalized on the wall included a picture of Um Harith's mother and aunt. Taken shortly before they died, the two women with prim expressions were sitting on a hard, wooden-backed sofa, their bodies cloaked in the severe black chador worn by pious Shiites.

Absent from the wall was any sign of Munaf, or the children born after him. Some nights, as the al-Sudanis ate under the gaze of the photographs, Munaf wished he could have squeezed onto that hard sofa at his grandmother's house when the photographer was preparing his shot. He hated being overlooked, but he slowly learned to exploit the advantages that his invisibility had presented to him. While Abu Harith ordered his eldest son to work with him each day in his shop, Munaf had the freedom to play soccer after school. He could skip chores and bring home mediocre grades knowing that if Abu Harith noticed, his scolding would be light, his disappointment fleeting.

And why should he excel at school? Why should he exert himself? He wasn't being raised in a meritocracy. The hierarchy of the family was a natural order reinforced by almost every interaction. On Friday afternoons Abu Harith would host his brothers for lunch after the weekly noon prayers. The men would sit around the cream-tiled floors of the family room, smoking and sipping tea. Younger children flitted in and out, serving fruit and nuts and clearing teacups, eavesdropping on the conversations about poetry and sports. Harith was a frequent topic of discussion as well, not Munaf. Their father would boast to the gathered relatives how Harith would be the first in their family to get a university degree and a prestigious job, and bring honor to the al-Sudanis.

Munaf never begrudged Harith the attention bestowed by their father. In fact, Munaf and his brothers often pitied the eldest. Once Harith was kicked out of university, Munaf could see how painful the humiliation was for his brother. There weren't many professions in Saddam City to choose from. Lower-ranking civil service jobs like teaching or nursing might be possible, but it was simply assumed that most young men would apprentice with their father or uncles. Either that, or they would join the growing ranks of the unemployed. In the al-Sudani family, Harith's ten siblings understood that Harith was expected to do something respectable, like become an engineer. The second son was expected to join their father in his small graphic design business. When it came to Munaf, there was no plan at all.

Even as a boy, the al-Sudani son had internalized the primal laws of Iraqi society. Predators prey on the weak, and under no circumstances should an Iraqi like him ever rely on the police, because they would as soon fuck you up, rape your mother, and leave your whole family for dead rather than help anyone, especially a poor Shiite like him.

Growing up in Saddam City, boys experienced a variety of casual violence, meted out as discipline at home or, even more toxic, from the sons of powerful families who mistreated the rest of the neighborhood with impunity. One reason that it was a blessing to be part of a large family was that there is strength in numbers. Brothers could be called on to defend one another against bullies and shakedowns. But even then, the tactic of strength in numbers could only be used in select circumstances.

When Munaf was in seventh grade, there was a boy in their neighborhood who was a year younger than him, a solitary boy named Salem who was an only child. Salem was easy prey for Hussein and his crew, a band of bullies led by the son of the neighborhood's informer, the man who reported to the mukhabarat, the secret police. No one could complain or fight back, because to do so raised the specter of jail, or worse, given the connections that Hussein's father had. So absolute was their reign of terror that the other kids called them the sheikhs of the street.

Almost every morning, the bullies would circle the streets between their homes and school and lie in wait for Salem. For the ten minutes it took to get to school, the pack would exhibit a finely calibrated kind of terror. They would surround the younger boy, trip and kick him, but to any passerby it would appear that they were all good friends. Salem had no escape from this daily torment. When he arrived at school, the principal would punish the boy for his disheveled appearance. Salem endured those whippings in silence as well, knowing the vengeance of the bullies would be greater if he told the principal what had happened to him. Some days, Salem feigned sickness to stay at the school building an extra hour to avoid the bullies on his way home. Other times, he tried to outrun them. But the next day, their hunt would start again.

Munaf felt sick to his stomach when he saw the bullies in action. But he knew that taking on the sheikhs was a foolhardy pursuit. Everyone knew that Hussein's father was likely responsible for the disappearance of the family whose house had become haunted. Munaf didn't want something like that to happen to his family, so he did what his father had always admonished him and his brothers to do. He kept his head down and stayed out of trouble. Munaf told himself that the bullying was good for Salem. Like all the boys in Saddam City, he, too, needed to learn fortitude to deal with life's brutality.

In a neighborhood with few televisions or other distractions, boys played outside until dinnertime. In springtime, the late-afternoon hours after school became as sacred to the neighborhood boys as the Ramadan fast, for that was the time when they could gather at the only empty lot within walking distance, the hard-packed sandy patch of land across the street from Munaf's uncle's house, and play soccer. While the boys changed out of their school clothes, they would press-gang their younger siblings who were playing nearby to pick up the rocks embedded in the earth. It was a hard task without any reward, as the field was reserved for the older boys only.

One afternoon, when Munaf was twelve, the street sheikhs were towering over Salem, forcing him to kneel on the ground in his clean school clothes and pick up stones with his mouth. Each time he dropped one, Hussein would cuff him behind the ear. Pick up the fucking stones, you dog, he barked at Salem. You have no life unless I say you do. You have no wishes except mine.

Munaf was standing on the far corner of the lot, but even from there he could see blood trickling from Salem's ear down his shirt collar. His eyes were squeezed shut and his face was wet with tears. Munaf walked on, pretending he hadn't noticed. Head down, eyes down, like his father always told him.

Suddenly, a piercing whistle came from down the lane. What in God's name are you playing at? You son of a whore!

A man in a dark-blue uniform appeared from nowhere, like a superhero in an American movie. His hands were the size of cinder blocks and his neck as wide as the trunk of an apple tree. He wore the clothes of a police officer and Munaf had never seen him in their neighborhood before.

The street sheikhs looked up, astonished, as the policeman picked up Hussein by the back of his shirt. His boot-bristle mustache quivered with anger. Shame on you! How dare you attack this child!

The man cuffed the young bully with a blow that sent him tumbling to the ground. Hussein landed heavily on his rear end. His face went white and then redder than a tomato. Who is your father? the police officer asked brusquely. Let's go tell him about what you have been up to.

The officer pulled the boy down the street toward the police station. The rest of the boys were dazed at the spectacle they had just witnessed.

Munaf remembers his heart pounding with excitement. The scene was the most heroic thing he had seen in his young life. Who would have ever believed that a policeman—an object of fear and derision—would be the vessel through which God revealed something wondrous? This stranger appeared like an angel and bent the law that had always been immutable in Saddam City.

Munaf couldn't put into words what he was feeling. Emotions coursed through his small body. His headache had disappeared. This was what justice looked like.

He didn't know the name of the man who had reordered his universe. But he knew he wanted to be like him. From that day on, he knew he wanted to become a policeman.

For years there had never been a time for Munaf to explain these feelings to his father. He didn't have the language to describe these thoughts, nor the gift of poetry like Harith. The last thing that Abu Harith would understand was a desire for his son to upset the social order.

Those born in a police state are provided no manual to navigate the ambiguities of power, where the distance between a joke and jail time can vary according to the mood of the informer. To Abu Harith the threat of arrest hovered constantly, like a swarm of wasps. His only method of inoculating his family against this peril was to mold them—and himself—into models of meekness. Obedience was his talisman.

The patriarch walked through life with his eyes down and head bowed, choosing whatever path he needed to avoid trouble. He gave his children names favored by Iraq's Sunni families, a subtle sign of obsequiousness to those authorities that would accost his children their whole lives. But to the families living in Saddam City, Abu Harith would explain his choices as venerating wise men from Islamic history. Politics was something to be avoided, especially as his government-issued identity card revealed him to be a double threat to the regime, given his permanent address and religious denomination.

For this reason, Abu Harith shunned the storied male bastion of the Arab world: neighborhood teahouses. The geography of Saddam City gave its residents few public spaces to mingle freely, so teahouses became one of the few places where men would gather after work, or spend their day if they were unemployed. Given how hard life had become, you didn't have to be political to rage about the cruelty of Saddam's regime in such convivial settings. Still, walls had ears in Iraq, just like every other totalitarian country. Merely listening to such rage left a man open to risk. If you stayed silent, you were vulnerable to being written up by the regime's informers. If you offered support for the government, you could face reprisals from the opposition groups.

When Abu Harith lost his printing business, he started writing freelance columns in pro-government newspapers and publishing poems under a pen name. As it was, meat was a scarce guest at the al-Sudani house—if his name was added to an informer's report about a teahouse conversation, even by mistake, Abu Harith would lose even that meager paycheck.

Even so, the al-Sudani family lived a relatively comfortable life, at least by the standards of their slum, and Munaf and his brothers saw their corner of Saddam City as a small piece of paradise. Abu Harith and his cousins had built the family's three-room unpainted cinder block home on a corner lot amid a warren of hard-packed dirt lanes where the extended family all lived and where children could play unsupervised at all hours. City sanitation workers never ventured into their neighborhood, but instead of cursing the government neglect, Abu Harith viewed the situation as a blessing. The family's goats munched on mounds of rubbish instead of feed that he didn't have the money to buy. Mothers recycled glass jars and bottles out of simple necessity. Siblings wore each other's hand-me-downs. Neighbors knew each other's names and each other's business.

But in 1999, the whole nation was suffering. Saddam's political feud with the Americans had brought Iraqis nothing but pain. Iraq was suffocating under international economic sanctions: children were dying from preventable diseases for a lack of medicine; the oil industry was rusting because there were no spare parts for machinery and infrastructure; and the once proud nation that boasted the birthplace of human civilization was on its knees.

In the absence of any political debate, Iraqis turned toward religion, and a revered Shiite leader, Ayatollah Mohammad Sadeq al-Sadr, became a beacon for the opposition, using his sermons to give voice to the belief of Iraqis from all faiths: Saddam Hussein was the reason for their suffering. Saddam had to go.

The ayatollah had been banned from television, so his representatives traveled around Iraq preaching directly to the faithful in mosques and spreading his message by word of mouth. Even Iraqis like the al-Sudanis, who steered clear of politics, understood the bravery it took to speak such things aloud.

The nation's impotent rage exploded in late February of that year, when state news agencies announced the death of Ayatollah al-Sadr and two of his sons. The terse statement gave no details, only stating that the men had already been buried before sundown, per Islamic rites.

Bewildered Shiites from the southern port city of Basra to Saddam City rushed to their mosques to pray for their spiritual leader's soul. Rumors spread through the crowds like an infection. Several months earlier, two of the ayatollah's closest aides had been assassinated. Three weeks before his own death, the ayatollah had been ordered by a government official to tone down his criticism. Even though they had not seen the body or known any details, the surging crowds were convinced that their leader had been killed by the regime.

In Saddam City, emotions were raw. Hundreds of men who had gathered for days at the Imam Hussein Mosque to pray for Ayatollah al-Sadr whipped themselves into a frenzy. A battle cry rang out through the mosque and into the streets. Death to Saddam, death to Saddam. The call brought hundreds more men out to the crumbling sidewalks and potholed streets, building to a crowd that became the largest popular demonstration since 1991, when the Americans launched the first Gulf War and called on Iraqis to topple the dictator themselves.

The al-Sudani boys didn't understand that trouble was brewing. It was a matter of luck that all the boys got home before the bullets started to fly.

Harith and his friend Ali had taken a bus that morning into Baghdad. Ali was just about to turn eighteen, and like all Iraqis, they had to register for mandatory army service at that age. However, both were finishing up their high school exams, hoping to put off their duty by winning a slot to study at university.

Ali wanted company as he navigated the bureaucracy of the military recruiter's office. Harith, with his deft use of words and manners with adults, was whom he asked. By late morning, the two friends were heading back to Saddam City on a city bus when they noticed from the bus window that truckloads of soldiers wearing helmets and carrying anti-riot gear were lining the main street leading into their district. Behind them on the four-lane road, large trucks carrying tanks and armored vehicles were clogging the main artery.

Munaf was at school when, around lunchtime, the principal announced that classes had been canceled. The boys were ordered to walk straight home without stopping. No soccer. No snacks at the corner market.

What they didn't know was this: less than three miles away, two battalions of Iraqi national guard and an army mechanized infantry unit had crawled into the district, tearing up the tarmac of the district's main street and occupying the central bridge that divided Saddam City from central Baghdad. Commanders had orders to restore calm. They and their men, the vast majority of whom were Sunnis from different parts of the country, knew failure was not an option. The deep boom of heavy artillery echoed through the narrow streets followed by the dull thuds of high-powered automatic rifles hitting their target, the people who had been praying at Imam Hussein Mosque.

The men and women who were inside the mosque tried to flee, but when they reached the streets they were torn apart by a hailstorm of bullets. The soldiers had set up a kill zone a dozen blocks wide and there was nowhere to hide. People lucky enough to escape with injuries didn't dare go to the hospital. The mukhabarat was scouring all of Saddam City's medical facilities for people with bullet wounds or shrapnel injuries. Either proved that the patient was a terrorist, they told the nurses, and they would be arrested and taken away.

For three days Abu Harith forbade his sons from leaving their house. The tanks had rolled back across the bridge, but the regime's message lingered like the stench of the bodies left in the streets by families too frightened to retrieve them. They could come for you at any time.

Abu Harith would not take the chance that one of his sons would be caught in the regime's dragnet. No one in the al-Sudani family had ever been arrested, even in the churning political foment of the 1980s. He wanted to keep it that way.

For Munaf, the afternoon rally in April 2003 began to thaw the emotions that had been frozen in the grip of his father's fear. With the dictator gone, the residents of eastern Baghdad renamed their district Sadr

City, in honor of their martyred ayatollah. The surge of hope lifted taboos ingrained for a generation. Some Iraqis cowered at the new era of uncertainty. Others jumped at opportunities created in the tumult. The borders were wide open for trading for the first time in a decade. Many families in Sadr City opened small businesses selling building materials, used cars, and appliances, commodities that for years had been available only on the black market but that every Iraqi needed and now wanted.

The transformation of the country also overturned the al-Sudanis' family dynamics. After Harith announced that he had flunked out of university, Munaf was no longer an afterthought for his father. In fact, for the first time in his life, the third-born son found himself the object of his father's attention. Abu Harith desperately wanted one of his other sons to cleanse what he considered a stain on the family honor, and nineteen-year-old Munaf had strong grades and much promise. He wasn't interested in a career in business or in academics, though. In 2004, shortly after he had graduated from high school, Munaf brought home a flyer advertising jobs with the new Iraqi police force. The slogan at the top of the paper was elegant and moving. Defend your country—Defend your people.

Munaf knew that mission was a worthy argument to give his father, the pitch that would allow him to follow his dream. A year later, he was enrolled in the new Iraqi police academy, and was on track to becoming one of the first cadets of the nation's young democracy.

Chapter 3
Breaking with the Past

Abrar al-Kubaisi was only fifteen when the U.S. military invaded Iraq. A petite teenager with slender arms and somber dark eyes, she had a nervous demeanor. Her family often told her how pretty she was, fragile as a porcelain doll. She never knew how to interpret that praise. She was one of three girls in the extended family, and she considered her older sister, Tasneem, the most beautiful girl she had ever met. Besides, the man who most often complimented her, her father, Mohammed, was blind. Better to look like a pumpkin instead of a bean stalk, her aunts, who could see, would tell her.

But Abrar never had a large appetite. Not even for her mother's famous dolma, rice-stuffed vegetables made by Iraqi housewives on special occasions. Nor for the baklava sweets that her father brought home after work at Baghdad University. There simply was no sign that her body would develop the curves that her aunts had, or that she would resemble one of the voluptuous actresses in the Egyptian films broadcast on Iraqi state television, the ones about star-crossed lovers, or the dashing hero who plucked a beautiful woman with wide hips and long wavy brown hair from her life of drudgery.

But even if she didn't look like a film star, Abrar had talents that her family was proud of. She was smart, and she was sure to earn a spot at the country's most venerable university, where her father was a professor of Islamic Studies and Arabic. As members of Baghdad's middle class, the al-Kubaisis lived a relatively comfortable existence. They were part of the city's educated elite and they benefited from their membership in Iraq's Sunni desert tribal networks, the political base that for so many years Saddam courted, coddled, and conspired with to stay in power.

The bubble in which Abrar was raised was so exclusive that she rarely met Iraqis who weren't part of her extended family, which consisted of uncles who were physicians, high-ranking civil servants, or professors like her father. She and her four siblings grew up playing in the garden of their two-story home in the leafy west Baghdad neighborhood of Amariyah, where a towering date palm tree bent toward the roof and the lemon tree blossomed in the spring. Her expectations were no different from those of other Iraqis born into the ruling class, the belief that their success and wealth were the natural order of things.

The al-Kubaisi family were members of Baghdad's urbane elite, successors to the city's ancient tradition across the Islamic world as a capital of books and learning. Mohammed al-Kubaisi, Abrar's father, refused to allow his own disability to break the branch of his own family tree that boasted scholarly ancestors stretching back to the end of the Ottoman Empire. It was a mark of great pride that he was the only blind professor at Baghdad University. One of his brothers practiced medicine, another was a high-ranking civil servant in the ministry of higher education. His oldest son was completing a Ph.D. in computer engineering. Abrar's mother taught at the girls' high school just down the street from their home.

As far as Abrar knew, all Baghdadi children were raised like she was, with a strong sense of heritage and respect. Her parents never discussed politics at home, and why should they? With Saddam in power, families like hers lived in safety and security. On her street in Amariyah, fathers wore suits and ties and worked at well-appointed offices. It was a mark of good manners and sophistication that no one discussed the fact that while her father was Sunni, like Saddam, her mother, Zainab, was Shiite. Islam, for the al-Kubaisi family, wasn't about how someone held their hands while they prayed, or the order in which they had visited certain holy sites, it was the rule book that set the cultural mores of their lives: what was proper behavior for little girls in public; the social obligations toward family and neighbors during the holiday season; and the standards of cleanliness that women had at home. The neighborhood wives, when they met at the market and gossiped, didn't quote Koranic verses or compare holy sermons. They commented on which tribal ties a family had, who had the most college degrees, or whether someone was well-connected enough to get a promotion. And they judged each other accordingly.

These were the intractable rules of their social network, the thread that bound them in the complicated compact between the ruler and his subjects. In the al-Kubaisis' experience, as long as one lived within these well-defined lanes, families like theirs could thrive. Even through the hardships of the 1990s, when Iraq was subject to crippling economic sanctions for the sins of its dictator in the aftermath of the Gulf War, while many Iraqis were pinched for basic goods like medicine or food, they managed. High-level members of Saddam's ruling party had set up vast smuggling networks, making a mockery of the sanctions placed against the regime, as most goods could be acquired if one had the right connections or enough money.

The al-Kubaisi home was usually full of such gifts, books published in the West, and blue jeans. Mohammed's brothers even made money in those hard times. Thanks to their jobs in the medical profession and their tribal ties in neighboring Jordan, they managed to sell hard-to-find medical supplies and drugs to their patients in Baghdad. After all, as they told Mohammed in the evenings at home, we have families to feed as well.

One of Mohammed's proudest boasts was that in 2002, his family was the first in Amariyah if not all of west Baghdad to obtain a personal computer. Despite its proud legacy, the Iraqi capital was hopelessly behind much of the world when it came to technology. Decades of U.S. and United Nations sanctions had prohibited the import of most electronics, something that consumers hated but from the regime's perspective was beneficial. Under Saddam's police state, anything electronic, especially telephones or computers, were sensitive, because of their potential to subvert the surveillance apparatus that the security forces had so carefully erected.

But Professor al-Kubaisi knew how important the coming computer age would be for his children. One morning just before lunch, a black GMC Suburban pulled up to their home, having been driven from Amman, the Jordanian capital, much earlier that morning.

The driver unloaded half a dozen oversized cardboard boxes that could barely fit through the al-Kubaisis' front door. Inside all the protective wrapping were the components needed for a desktop computer.

Professor al-Kubaisi wasn't afraid of any political repercussions after his purchase. Families of his standing didn't need to be. His oldest son, Mustafa, had started studying computer science, and assembling the computer was a skill sure to put him at the top of his class. As Abrar watched her older brother construct the motherboard and the hard drive with his screwdrivers and wire clips, he could tell right away that technology also held a fascination for her. When he was finished, they placed the machine and its screen on a desk in the center of the living room—in pride of place in the house. Although the computer had been bought with only Mustafa in mind, Abrar convinced her father to let her use it, too. She spent hours exploring the amazing contraption and begged Mustafa to teach her how to type. She earned more time on the computer by offering to type up Mustafa's lecture materials for his classes at university.

Luckily, Baghdadis of her parents' generation and status encouraged their daughters to be as educated as their sons. Her older sister, Tasneem, couldn't care less about studying—she wanted nothing more than to be married. But everyone understood that Abrar had inherited the family's talent for learning.

Once the Americans started bombing Baghdad in 2003, the al-Kubaisis' confidence quickly turned to fear. No one from the immediate family joined the street rallies erupting around town. For the first time in his life, Mohammed forbade his children from leaving the house.

As they huddled inside evening after evening through the spring of 2003, pictures streamed on television screens across the Arab world of the crowds celebrating their freedom from Saddam. Um Mustafa, Abrar's mother, remembers the hunted, bewildered look of the former dictator when he was captured by the American troops. While Mohammed couldn't see these images, he was filled with foreboding about the changing realities.

He would quote another well-known saying—that Iraqi blood runs hot. Professor al-Kubaisi was sure that this emotional fever would descend like one of Baghdad's notorious dust storms, spreading noxious clouds of sand and erasing the landmarks of his family's life. It didn't take long for him to be proven right.

By 2004, the pent-up misery and repression of Iraq's majority Shiites boiled over into rage against the Sunni community who had either abetted the former regime or passively benefited from it. The cosmopolitan, urbane districts of Baghdad, streets where famous poets and actors held court, weren't spared, either. The contagion of sectarianism spread so rapidly that many families found themselves on specific sides of battle lines drawn by virtue of demographic reality, with Shiite militias desperate to wreak vengeance for the sins of the past and Sunni militias trying to hold on to some semblance of the life they had long experienced. Amariyah, the al-Kubaisis' neighborhood, became a flash point in that war.

Abrar watched the street battles between militias play out on the evening news, safe within the family's villa. But early one morning she heard an unusual commotion coming from the house next door. Her neighbors were outside unloading a pickup truck full of household goods, a mattress, a refrigerator, and bulging hemp sacks bundled in twine and rope. A young couple she had never seen before were trying to hush a crying infant, while the mother herself was weeping.

Later in the day, Abrar knocked on her neighbors' door with a cake, an excuse to visit and find out just what was going on. The young wife, Noor, told Abrar the story. She had been living happily with her husband, Ala'a, in the southeastern district of Baghdad, Jdeideh, a place full of newly built apartments for young, upwardly mobile families like theirs. But then one fine day Shiite death squads dressed in black shirts and pants swept into the neighborhood, taking up guard posts on busy intersections and checkpoints leading in and out of the area. They spoke with an accent of Iraq's southern Shiite heartland, the same as spoken in nearby Sadr City. The militants demanded protection money from Shiite shopkeepers and warned Sunnis to leave. Groups of men in the militia's unofficial uniform would stand at the bus station in the evening waiting for men to return home from work, pulling people aside like gangsters and demanding to see their government identity cards. "We don't want rats like you in our neighborhood," they'd say to those with Sunni names.

The message was clear, but Ala'a didn't want to be run out of his home. Less than a year earlier, they had finished decorating it with the dowry from their wedding. They had a sunny room for their baby. Maybe, he thought, this madness will pass.

But then the violence came straight into their lives. The couple awoke to the sound of men running up the apartment building stairwell, shouting all sorts of foul language. Next, they heard gunshots. Their neighbor, who was Sunni as well, was shot dead in his doorway. His body lay on the landing, blood spreading in a large pool around him, until morning. Neither the police nor paramedics would risk their lives to respond to a call in a district patrolled by the militia, called the Mahdi Army. Ala'a and Noor spent the evening packing as many of their belongings as they could and drove directly to his uncle, Abrar's neighbor.

As months marched on, in turn Amariyah's Shiite families were forced to flee their homes after receiving their own death threats. More and more Sunni refugees from neighborhoods on the east side of the river moved into the abandoned properties. Ala'a, who was in need of money, started driving Abrar and her father to Baghdad University each day, escorting them past the phalanx of checkpoints that choked the city's traffic. When they took the Jadriya Bridge across the mud-colored Tigris River, he would point out corpses floating in the water. Dead bodies had replaced the ducks that normally swam in those deep waters. The concrete barriers that had grown in the government district known as the Green Zone and the hard demeanor worn by the campus guards at the university reminded Abrar that everything she had considered decent and normal had changed.

When she stops and considers it, which isn't very often, Abrar believes that the moment she decided to act was when she watched, riveted on the sofa in her parents' living room, a young Iraqi woman, someone just like her, as she told millions of television viewers that she had been raped.

That evening, Um Mustafa had wanted to watch a cooking show. Professor al-Kubaisi had wanted to listen to his favorite sheikh's call-in program that focused on the loss of morality in the Arab world. But Abrar, who had finished her homework for her classes at Baghdad University, prevailed in her argument to control the remote. She tuned in to Al Jazeera, which by the start of the Iraq War had become the channel that millions of Arab Sunnis tuned in to each day to hear the latest developments on subjects they already knew plenty about—the perfidy of the American occupation of Iraq, the murderous intention of the nation's Shiite militias, and the rationale for resistance against U.S. forces and the Shiite-dominated government that agreed to the foreign military presence in their ancient country.

As they huddled in their homes each evening, every Iraqi household was desperate either for entertainment to escape the escalating chaos outside, or for news that might provide enough information to make sense of it. Television was the answer for both impulses. On that winter night, Al Jazeera broadcast an interview with a young woman, a newlywed from a Baghdad neighborhood not far from Abrar's. Twenty-year-old Sabrine al-Janabi wore a pink-and-white-patterned headscarf. Her body was completely covered in a shapeless black garment and she was reclining on a bed covered with a colorful plush blanket. Her face was smudged with tears and mascara and her dark brown eyes were darting as if in fear. Sabrine spoke in a soft voice with a distinct Iraqi accent that halted and broke repeatedly as she told tens of millions of viewers how three Iraqi national policemen arrested, assaulted, and raped her.

According to Sabrine's account, her ordeal had started two days earlier. Her husband was out, and she was alone, when the Iraqi police unit broke into their home announcing that they were on a counterterrorism operation. The police told her that they were searching the street for insurgents and illegal weapons. They then bound her hands and shoved her into a car with a dozen other detainees from the Amil neighborhood, a Sunni area like Abrar's. Sabrine was the only woman in the police vehicle and she hadn't been told why she was being taken away. When the police arrived at their garrison and started processing the detainees, they accused Sabrine of providing food for Sunni fighters who had been battling the Shiite-dominated national police forces like them.

At that point, she said, they shoved her into a small annex, a room with a bed and a Kalashnikov machine gun against the wall. That was where she said that the first officer raped her while he covered her mouth to muffle her screams.

"I told him, 'Please—by your father and mother—let me go.' He said, 'No, no—by my mother's soul I'll let you go—but on one condition, you give me one single thing.' I said, 'What?' He said, '[I want] to rape you.' I told him, 'No—I can't.'"

Another officer came in and told the first man to leave. Sabrine then told her interviewer she tried to defend herself: "I swear on the Koran and the Prophet Muhammad, I am not that kind of woman," she said. She next describes being beaten by the second officer with a black water hose.

Sabrine then broke down, telling the reporter she was unable to continue.

In the al-Kubaisi home, the family was struck silent by the report. Then Abrar started sobbing, unable to stop for the additional twenty minutes that remained of the hour-long news program. She wasn't alone in her grief.

Sabrine's shocking interview transfixed Iraq's Sunni community and increased the already loud drumbeat of outrage. In the days prior to the interview, Al Jazeera and many other Arab news stations catering to a Sunni audience had been reporting about the verdict against a U.S. Marine who had raped a fourteen-year-old Iraqi girl from south Baghdad. He and his platoon killed her parents and then burned all their bodies in hopes of not getting caught.

That evening, Al Jazeera fed into this anger by featuring a steady stream of politicians from Sunni communities. One by one they demanded justice, calling the allegation a defilement of their community's honor.

To a nation already numb from an unceasing number of grisly murders, the rape allegations exposed perhaps the one remaining taboo for society, a crime so horrible that speaking of it publicly paralyzed politics for days to come.

Iraqi Shiite leaders lined up to condemn the woman, calling her a liar and provocateur. They claimed that Sabrine fabricated the story to undermine the government of the new prime minister, Nouri al-Maliki, for the lack of results in curbing escalating terror attacks. Sunni politicians countered with numerous allegations of rape among their constituents. Anger churned into a further frenzy when Prime Minister al-Maliki issued a statement promising to investigate the police unit that Sabrine accused of wrongdoing, only to have his office issue a contradictory statement hours later calling those policemen heroes and Sabrine's accusations false. A few days later, the Interior Ministry announced that they had issued an arrest warrant for the young woman and accused her of polygamy and supporting terrorism.

In Iraq's combustible atmosphere, where rival sectarian and ethnic militias were cleansing one another out of whole neighborhoods, the development set off a new firestorm.

At least six groups, including Al Qaeda in Iraq, called for revenge attacks in the name of restoring the honor of Sabrine. Iraqi Sunni politicians made the young woman a rallying cry for justice, calling her the "sister of the nation." The young woman never again appeared on television, nor gave another interview. But the young women who were studying chemistry with Abrar at Baghdad University couldn't stop talking about the case.

They knew, even if the nation's leaders did not, that no Iraqi woman under any circumstances would publicly admit that she was raped—or falsely claim that crime. The risks are simply too great. First, there is the risk of being shunned socially for the shame that such a crime brings to the woman. No one would ever agree to marry her, and her own life could be forfeit. Then there is the risk of sparking a chain of retaliation and revenge killings between tribes. In Iraq's deteriorating state, tribal courts and not law courts were the venue to which the nation's large families turned for satisfaction and justice, and rape carried a uniquely steep jiziya, or blood price, to be paid by the perpetrator.

Abrar couldn't sleep, stuck in the horror of Sabrine's life. So, while her family were in bed, she turned on the computer. There, in the reflected blue light of the monitor, she sat and poured out her feelings to a group that shared her anger.

The news of Sabrine's defilement hit Professor al-Kubaisi deeply as well. He felt like he could no longer trust the basic humanity of his fellow Iraqis. The following day, he and his wife sat Abrar down at the dining room table determined to give her some bad news. Yes, they told her, we raised you to get a Ph.D. like your father. Yes, of course we always expected you to use your education to get a good job. But, God help us, how can we live with ourselves if something would happen to you at the campus, or at the checkpoints?

Mohammed broke down in tears, surprised at the force of his emotions. He couldn't understand what the world was coming to. Abrar also started to cry and pleaded with her parents to allow her to stay in university. The one source of stability that she had left was her studies, and she was determined to pursue them with all her strength.

When she was in the university laboratories, Abrar explained, she finally managed to shut out the chaos of the world. Her parents finally relented, allowing her to remain in her classes.

Abrar spent as much time as possible at the laboratories. As she told her parents, she felt at peace in the antiseptic coolness of the workspace. Instead of the lawlessness and chaos of Baghdad's streets, in the research lab she found precision and order. What Abrar didn't mention to her parents was that she was slowly finding a channel for her budding political feelings. Her work after hours amid the rows of narrow-necked beakers and Bunsen burners taught her just how easy it was to create an explosion. Mixing the right formula of chemicals might lead to a lifesaving medicine, yet just a milliliter more or less of a single liquid could lead to combustion.

Chapter 4
Return of the Exiles

In the spring of 2003, while the al-Sudani brothers were celebrating the arrival of the Americans in Firdos Square and Abrar nervously followed the news on TV, Abu Ali al-Basri drove into Baghdad after almost two decades in exile. He was ecstatic that his homeland was finally free of its dictator.

Just two weeks before the U.S. tanks rolled into the Iraqi capital, Abu Ali had been in a Swedish hospital helping his wife give birth to their fourth child. After so many years of making his work in the anti-Saddam opposition the priority over his family, a life full of false passports, long weeks in the mountains of Kurdish-controlled northern Iraq, in safe houses hidden in the warrens of Damascus's old city or Tehran, Abu Ali had given in to his wife's pleadings to make a stable life for her and their family in Sweden. He had set aside the fieldwork demanded by his opposition party, Dawa, and ran a social program aiding the large group of Iraqi immigrants that Sweden had welcomed in the 1990s.

That, however, didn't mean he had retired from the heady world of Iraqi politics. Abu Ali knew, as did the leaders of Iraq's top opposition parties, that shortly after 9/11 the Bush administration had started planning to invade Iraq. For his long time in exile, he had not learned to speak English and he wasn't part of the exile cliques selected by the Americans in their early meetings held in London; Amman, Jordan; and Washington for advice on how to prepare the ground for the U.S. forces, rewrite Iraq's constitution, and lead a transitional government. Unlike those exiles, Abu Ali hadn't cultivated close contacts within the Central Intelligence Agency (CIA) or State Department, and he wasn't interested in jostling with them in the high-stakes political games over who would take control over the country's resources or political process. What he wanted most of all was to return to his homeland and, after reconnecting with the family he had not seen in thirty years and smelling the orange blossoms flowering that spring, help put right the injustices that they and millions of other Iraqis had suffered at the hands of Saddam's mukhabarat.

One night in 1979, when he was a teenager living with his family in the Al Saniyaa neighborhood in Basra, Abu Ali's father, a soft-spoken, lean man with a wispy beard, was stretched out in the majlis, the room used for family gatherings and for guests, his back resting on the red and yellow wool cushions that lined three walls of the room. Dinner had been cleared away and his father was chatting with two of his brothers-in-law, while his mother was down the hall in the kitchen cleaning the dishes. The thick adobe brick walls of the home built by Abu Ali's grandfather kept the room pleasantly warm past sundown.

Young Abu Ali sat with the men, sipping tea from the small, narrow-necked glasses that his mother had received as a wedding present, and which the children were never allowed to touch. But, as the date approached for Abu Ali to register for mandatory military service, his mother had started serving him tea from one of the delicate treasures as well.

The sweet fragrance of orange blossoms wafted through the open window at the other end of the room, where Ali's younger brothers and sisters were trying to coax the family's songbird to sing. Their laughter must have covered the rustling noises outside because the first sign that something was wrong was the sound of fists pounding on the front door.

Come out, a deep voice shouted from the courtyard. Come out, you bastard.

The children froze, startled by the sudden violence. Everyone in the room turned to Abu Ali's father. His eyes were bright but glassy. Ali thought of a rabbit he had once trapped while visiting a relative's farm. More fist-pounding on the door.

You son of a whore. I said open the door, the voice shouted again.

Before anyone could move, a gunshot echoed like a thunderclap, and men wearing the khaki uniform of the mukhabarat stepped through the broken door and into the hallway.

They didn't have to identify themselves. Everyone understood who they were. What filled the family with dread was not knowing the reason why they were there.

Abu Ali's father was a bricklayer who spent every night at home with his family and who attended daily prayers at the mosque. At the start of the bloody Iran-Iraq War that year, piety among Iraq's Shiite majority was tantamount to treason. In the eyes of Saddam's mukhabarat, only Iraqis who were Sunnis like their leader and members of the ruling Baath Party were trustworthy citizens. Everyone else were potentially

Communist Party supporters and Shiites—enemies of the state.

A tall, spindly man with a beret and a star on his shoulder board pushed his way into the family's majlis. The officer pointed at Abu Ali's father, and without saying a word three men grabbed him. One of the men kicked his father in the kidneys while twisting his arms behind his back as two others pulled him toward the street.

Abu Ali's father grabbed the doorjamb, desperate to stave off the inevitable. Why are you taking me? he cried. What have I done? Why are you treating me like an animal?

A policeman pounded Abu Ali's father's hands with his rifle butt. I beg you to let me go, he sobbed. On the souls of my sons I have done nothing wrong. Abu Ali could hear his mother's high-pitched wailing from the kitchen and felt the blood rush to his head. Without thinking he jumped on the officer who was breaking his father's bones. Leave us! Get away from my father!

The officer laughed at the short boy whose arms were like twigs and who weighed little more than a wet dog. He pivoted his weight and swung the rifle butt directly at Abu Ali's head. Motherfucker, get off me, he growled.

The blow knocked Ali to the wall. Two other men pulled his father through the alley to the waiting vehicles and shoved him into the trunk of one of the cars. The drivers revved their engines and the cars tore off.

By morning, the only physical evidence that his father had ever existed was his half-filled tea glass he had been drinking from and his photo hanging in the front hallway. Abu Ali and his mother visited every police station in Basra every day for a week. No one had any record of her husband's arrest. No one could tell her who had ordered his detention. No one admitted to holding him.

After his father's disappearance, men watched their street and children at school stopped talking to Abu Ali and his siblings. One night, street thugs threw a Molotov cocktail through their kitchen window, igniting the canister of cooking gas. The family barely escaped. That was when Abu Ali's mother told her son it was time to flee. There would be no future for him in Basra as the eldest son of a presumed enemy of the state. The nation was at war with Iran and she was afraid that when he reported for the draft he would be arrested. She wanted her oldest to try to find a life for himself, but the only way to do that was to drop his father's name. They know who you are, she said. This means you no longer can be who you were.

Several days later Abu Ali took a bundle of food in a canvas knapsack and stowed away on a slow-moving train north to Baghdad, his first-ever trip to the Iraqi capital.

As a child, Abu Ali had played a game with his younger brothers, a version of a guessing game that they called Who Am I. They didn't have a yard to play in, and they couldn't afford a soccer ball, the most popular pastime for boys their age. Instead, he and his brothers would hang around the neighborhood's corner shop where housewives would buy eggs, olive oil, and cooking gas.

When a stranger walked by, the brothers would take turns guessing where he was from and what profession he was in. A man wearing a suit and tie with ink-stained fingers was likely a lawyer. A woman wearing a chador and speaking lilting accented Arabic could be a neighbor's widowed mother visiting from Lebanon. The person with stained trousers and dirt under his fingers probably was a bricklayer, like their father. The fat man with the gold watch: the district's stingy real estate speculator who also worked as an informer for the regime.

In Baghdad, scared and alone for the first time in his life, Abu Ali made full use of the detective skills he had honed as a child. In a country full of informers, he didn't want anyone to remember him, so he wore anonymity as a tight-fitting suit. The easiest place to do this was Karkh, the neighborhood near Baghdad's grand railway station, the former terminus for the Orient Express, where the surrounding streets were crawling with police, but also with strangers. Hundreds of Iraqis disembarked from the trains each week looking for work, escaping the draft and arrest warrants, like Abu Ali, or stoking political fantasies of bringing down the regime.

Ali stocked shelves in warehouses, cleaned soot from railway engines, and sold newspapers. By night, he relied on the kindness of strangers, men who would share bowls of white beans, bread, and cucumbers, and a clean corner in which to sleep. He soon found out that this network of men who opened their warehouses and garages at night were part of a political underground, Iraqi Shiites who decided they would do anything to stop the new Sunni dictator, or Communists who promised workers' rights after Saddam's downfall. That's how the young man from Basra first heard of the Islamic Dawa party. Dawa followed the teachings of the Shiite clerics in Najaf, the Iraqi shrine city, and sought to overthrow Saddam. They cared about justice for the poor people of Iraq, families like Abu Ali's. Soon, he gained a nickname from those men who, like him, were trying to stay out of the mukhabarat's web of surveillance. They called him "Abu Isim"—the father of names.

Then, life shifted again. In 1980, Dawa's leader, Muhammad Baqir al-Sadr, was executed by the regime. That night, Abu Ali and other activists papered the capital with posters memorializing their leader. A year later, with help from the Iranian government, Dawa launched a suicide attack against the Iraqi embassy in Beirut, killing the ambassador and sixty other Iraqis. In 1982, medical students at Iraq's most venerable university tried and failed to assassinate Saddam Hussein in Baghdad. The plot details remain murky, but the mukhabarat blamed Dawa, turning the organization into Saddam's public enemy Number One.

Abu Ali and the other cell members in the capital knew a crackdown was coming. Those who could started to plan routes out of the country.

One night, after midnight, the mukhabarat raided the safe house where Abu Ali was sleeping. The armed men were calling out his birth name—Abdel Khalid—not the alias he had been using. He managed to slip into a shallow hole dug into the earthen kitchen floor and cover himself with a woven rug before the officers came into the room. Luckily, it didn't occur to them to look under the carpet.

The close call spooked him more than he cared to admit. Abu Ali decided it was time to flee again, this time out of the country.

As a man of military age, Abu Ali could not legally apply for a passport without the risk of arrest. Without a passport, he had no chance of crossing the border. His Dawa contacts told him of a forger who lived a couple of streets over, someone who could get him travel documents. That news made Ali nervous. A man like that could sell him fake papers and then the next day sell that information to Iraqi intelligence. He feared he was walking into a trap. But what choice did he have? The mukhabarat were hunting him. Their grip, he thought, was closing fast.

Abu Ali set up a stakeout, watching the forger's shop for two days and learning the normal rhythms of the man's work. He wanted to use this knowledge to inform his plan of approach, so he would sense if something was wrong. The street looked a lot like his own did in Basra, rows of mildewed, two-story wooden homes mixed with weathered, concrete buildings pockmarked by dust storms. Wilted, deteriorating vegetables rotted on the street corners. Telephone and electricity lines tangled in complicated webs across storefronts. Neighborhood wives visited the corner store for bread, eggs, and tea. Children walked to school.

In the afternoon of the second day, he decided it was time to make his move. He knocked on an iron door on the home halfway down the street. A teenage boy opened it and showed Abu Ali to his father's workshop, saying his father would be home shortly.

Abu Ali stared at the wooden workbench and the rows of metal tools hanging on pegs. He hadn't slept for two nights and his nerves were alive with anxiety. This is a perfect setup, he thought. For all I know these people have already called the police. But Abu Ali didn't run. He had successfully evaded arrest for three years. Whatever happened next, he thought to himself, was his destiny, something he could not change or avoid.

After several more minutes of waiting, the door opened again. To Ali's surprise a young woman walked into the room, the daughter of one of the senior Dawa members in the neighborhood. Her name was Sarah. She laughed when she saw his surprise. Don't worry, I'm not here to arrest you, she told Ali. I'm here for a passport, too.

Two nights earlier, she said, the mukhabarat had arrested her father. As an unmarried young woman without a guardian at home, she couldn't live alone. With the political activities the family had been involved in, she couldn't safely stay in Baghdad, either.

There, in the dim light of the workroom, Abu Ali and Sarah forged a plan. Both of them needed to leave Baghdad and it would be safer to travel together. Each of them knew what going to prison entailed: physical torture, with the added brutality of rape if the prisoner was a woman. Sarah was determined to escape that gruesome fate and saw Abu Ali as an unlikely answer to her prayers. Abu Ali saw the same stroke of Providence. Most Iraqi men were married by the time they were twenty-one, paired with a wife through the matchmaking skills of their mothers and aunts. He hadn't seen his mother for almost three years and didn't know if he would ever see her again. He had never thought that a suitable candidate would walk down the same path that life cursed him with. After talking for only a few moments together the two had made up their minds. When the forger returned to the workshop, Abu Ali asked him to make their fake passports as if they were husband and wife.

While the forger worked on the documents, Abu Ali slipped away to Sarah's uncle's house, where she had been staying since her father's arrest. With his approval, the couple would be formally married in the Islamic tradition, removing any stain or scandal from their journey.

Ali secured train tickets to Mosul and the couple left Baghdad the next day. Sarah wore the black chador favored by conservative Shiites. At the start of their journey, she needed to be invisible, and the garment was a perfect disguise. Scared to peer around the veil, Sarah missed out on the scenery during her first trip outside Baghdad. As the train chugged northward, they passed the wheat fields of Diyala Province, the mountains that lead to Iran, and, when they reached Mosul, the famous leaning prayer tower known for centuries by its nickname, "the hunchback."

In Mosul, the couple didn't dare rest. Their destination, the Turkish border, was another day's journey by bus, through green hills full of spring wildflowers and thin tendrils of wheat standing tall in fields. The bus struggled up the mountain range that marked the Iraqi frontier. Shadows were growing longer, and Abu Ali prayed that the bus wouldn't break down. He wasn't sure that his and Sarah's documents would withstand the scrutiny of intelligence agents who would be sent to inspect anything unusual, like a packed bus on the side of the highway.

Their luck got worse. A few miles before the border, the skies opened with a cold, drenching rain. The bus belched to a halt a few hundred yards from the border crossing in a soaking storm.

Abu Ali guided Sarah forward through the rain, carrying their bags in one hand and holding her arm in the other. Klieg lights lit up broad patches of the border crossing where cars and trucks waited to pass.

The pedestrian queue, where they were heading, snaked deep into the darkness leading to a guard house about a hundred sixty-five yards away. As they inched forward in line, Abu Ali could see three men inside, armed, like all Iraqi officers, with pistols on their hips. The men took long minutes scrutinizing each person as they presented their papers in silence. Cigarettes dangled from their fingers as they thumbed through passport pages. The officer in charge, a bear of a man whose belly strained the buttons on his uniform, had a black brush mustache, just like Saddam's.

Ali and Sarah were shivering, wet to their skin. They tried not to think of the family members they were leaving behind. They told themselves that life might return to normal if they could just make it out of Iraq alive. But they had to get across the border first.

When they finally reached the guard booth, Abu Ali reminded himself to be steady and calm. He handed his passport across to the officers while Sarah stood two steps behind him, water puddling around her feet. The walls of the hut were covered with brown vinyl paper in a wood-grain pattern. Three men were squatting in a corner, handcuffed and blindfolded, with a thicket of welts and bruises on their faces. The silence was oppressive. Icy raindrops fell from the eaves straight down their backs. Ali was determined not to show fear.

Abu Ali and Sarah's passports were in the names of a real married couple from the Karkh neighborhood where the forger lived. He had never met them and didn't know them. What mattered was that the man and his wife had never applied for a passport. The IDs they carried should be clean as far as the border control guards were concerned. The only person Abu Ali was afraid of was the intelligence official sitting behind the desk. The three stars on his uniform revealed his status as a major. He had the power to detain someone just because he didn't like how they looked, or the color of their hair.

Salaam wa Alaikum, Abu Ali told the major. The officer ignored his greeting. He took a cursory look at the passport and then focused a cold stare on Sarah. Passport, he barked at her. Where are your papers?

Sarah's body shuddered with fear. She fought the burning instinct to turn around and run back into the darkness, to some semblance of safety. Her hands were invisible in the voluminous folds of the chador. Abu Ali saw he needed to play the part of a controlling Iraqi husband. He reached over and yanked at her arms to free them from the tent-like fabric so she could locate her passport and hand it over. Sarah remained rigid and silent, eyes on the ground.

Where the fuck are your papers? the major shouted again.

His tone that had started out hostile became even more aggressive. Like a wolf, he smelled their vulnerability. The major stood up so suddenly from behind the desk that his balding bare head hit the lightbulb hanging above.

Why does this whore refuse my order? he barked at Abu Ali. Can't you control her? What kind of man are you? He looked down again at Ali's passport, this time studying the information more intently.

Hand me your passport, whore, or my men will take you out back and take that and more from you, the officer growled.

Sarah started crying. Ali finally managed to find her hands through the chador. He pried her fingers apart and gave the passport to the major as the other two guards grabbed them both.

Now the officer was agitated. He fired off questions as quickly as a Kalashnikov rifle. Where did they live? What was the name of their paternal uncle? How did they afford their trip to Turkey? How long would they be gone?

Ali couldn't tell whether the water dripping from him was panicked sweat or the rain. He knew that their lives would be over if the guards dragged them away. He tried to keep calm and kept his eyes focused on the fabric just above the major's right knee. God was generous, he reminded himself, and would help him survive.

Just at that moment, a commotion erupted outside the guard house among a crowd of people entering Iraq from the Turkish side of the border. Women were shouting and wailing and the Iraqi guards were threatening to open fire.

The major ordered his men to investigate the ruckus. The guards left Abu Ali and Sarah and rushed out into the rain. The major lit a fresh cigarette. As he threw the lighter back down on his desk, he picked up the passports and threw them at Abu Ali. The movement came as such a surprise that Ali was not fast enough to catch them. The papers landed in a puddle of water that had formed around the hem of Sarah's chador.

Get the fuck out of here, the major yelled at the couple, waving his cigarette at them like he was swatting a fly.

Abu Ali took Sarah's hand and walked as fast as he could to the bus waiting to take travelers across the no-man's-land to Turkey. Don't look back, he told his wife as the bus pulled away. They can't hurt us now.

Four days later, Abu Ali and Sarah arrived in Damascus. They went from the central bus station to the area near what used to be the Jewish Quarter of the old city. They were desperate for food and a place to sleep and went to the only place they knew: the Dawa Party office. Again, destiny seemed to smile on the couple. While they were waiting at the office, the head of the opposition group in Syria arrived, a towering, lanky former Iraqi civil servant turned religious firebrand named Nouri al-Maliki.

After a few cups of tea, al-Maliki took a liking to this soft-spoken man who had an encyclopedic knowledge of the Baghdad underground. Before the afternoon was over, he had hired Abu Ali to work for him. Before long, Ali went from being in charge of numbingly bureaucratic tasks such as filling out paperwork to extend the residency permits for Dawa members in Syria to helping with more sensitive tasks.

One of those was implementing al-Maliki's plan to train Dawa operatives in a camp inside Iran, where the Shiite government was at war with Saddam and was eager to aid their Iraqi coreligionists against the dictator. The Iranian logistics and funds boosted all Iraqi opposition parties at a time when the Americans were supporting the Iraqi dictator.

Iraq's status as the world's third-largest oil producer gave Saddam an abundant war chest to pay for loyalty. Dissenters died under suspicious circumstances both at home and in Europe, where the Iraqi opposition had memorized the list of countries that were unsafe. Greece was dangerous, given the amount of corruption in the police forces there. Jordan was even risker because every Arab knew that the king kept his throne because of the heavy economic subsidies that Iraq paid him. The Arab method of relationship building was straightforward and transactional: If Saddam told the Jordanians not to let an Iraqi political dissident into the country, they would follow that order. If Saddam wanted someone assassinated in Jordan, they would turn a blind eye.

Abu Ali spent ten years going between Syria and Iran, assiduously building the party's counterintelligence department, which cultivated agents and sources inside the regime to collect information about their enemy and tracked the movements of Saddam's agents who were trying to infiltrate the party. The intelligence network boosted Abu Ali's reputation as he helped keep opposition leaders safe.

It was a strange and unexpected new life. Sarah gave birth to their first child, then two more, and worked to fashion a life among the other Iraqis in exile in Damascus and Tehran. News from inside Iraq was grim. After their escape, two of Sarah's sisters had been arrested. They later learned they had been raped and killed in prison. The uncle who had agreed to their marriage also died in an Iraqi jail.

In 1991, when the Gulf War began, Abu Ali and the rest of the Iraqi opposition were elated. American troops pounded Saddam's forces, the regime was in disarray, and U.S. president George H. W. Bush told the Iraqis to rise up against the dictator, saying they should "take matters into their own hands to force Saddam Hussein, the dictator, to step aside."

Dawa supporters and thousands of other Shiites did just that. They took over the town halls and barracks that Saddam's forces and officials had abandoned in their chaotic retreat from U.S. soldiers. They hunted down the men who had oppressed them for so long. But then everything went horribly wrong. American troops pulled out of the country, leaving Saddam strong enough for vengeance. Instead of having the opportunity to build a new nation, the Shiite opposition was massacred.

Abu Ali couldn't shake his melancholy. Had he wasted his life on a naïve dream of freedom? The price he had paid for spending years helping the opposition against Saddam seemed worthless, especially when, soon after the disastrous uprising, word reached him that his mother had died.

Grief-stricken and exhausted, Abu Ali, for the first time since he and Sarah had fled Iraq, contemplated doing something different in his life. Sarah begged Ali to get their family to a safer place, to leave his work behind and try to live a quiet existence.

He complied. Like thousands of other families from the opposition, they formally applied for refugee status. Within a year, they had resettled in Sweden, joining an exodus of intellectuals, businessmen, and more simple men like him who had no future in their homeland.

Sarah struggled with the winter weather, months of darkness, and bone-aching cold. But she reveled in Swedish order and routine. For the first time in a decade they had a semblance of normalcy. Abu Ali understood that the move was practical and necessary, like the secondhand winter coat he had received at the welfare center when they arrived in the frigid nation. But he was never entirely comfortable. He learned Swedish, just like his children, and he started an after-school program for immigrant teens. But Abu Ali stayed obsessed with Iraqi politics.

Over dinner each night, Abu Ali saw the joy on Sarah's face when they all sat down together. But, if he was honest, he didn't feel at home. In Sweden, there was no reason to watch your back and no reason to live under an assumed name. People were free and trusted their government. But back in Iraq, his nation was still suffering.

Saddam Hussein once joked that he killed traitors before they were even sure they would betray him. Iraqis knew the dictator wasn't exaggerating. By the time Saddam was overthrown in 2003, Iraqis could not name the various agencies, departments, and security forces working by any means necessary to help root out his enemies from the population. A CIA report from 1979 runs more than forty pages long in its descriptions of the names and duties of these different agencies. The overlapping maze of security services resembled the bureaucratic equivalent of an M. C. Escher drawing that, like the artist's staircases, existed without logical foundations, climbing from shadowy recesses into a web of pathways that bend back in an infinite loop of terror.

On paper, these edifices appeared solid and functional. Yet when the Americans unseated Saddam, those structures were ready to topple under the combined weight of their own corrupt history.

The American officials planning the invasion decided well in advance that, once their military had toppled Saddam, they would have to replace the entire leadership of the Iraqi armed forces and security services. In May 2004, the man who served as the head of the American governing authority in Iraq, Paul Bremer, made this decision official by signing Coalition Provincial Authority Decrees No. 2 and No. 3, dismissing thousands of Iraqi military men and intelligence agents due to their ties to the former regime.

In their place, the Americans had plans to build a new army, intelligence service, and police force, and they had several candidates lined up to take command. Abu Ali's name wasn't on those lists. In fact, the Americans had no idea who Abu Ali was.

From the CIA's perspective, the only man in the running for the job to head up the new, one-billion-dollar national intelligence service was a retired Iraqi general who had been on the agency's payroll for around twenty years and had become the drinking buddy of a generation of senior officials. Mohammed al-Shahwani, a retired Iraqi special forces commander and former Olympic wrestler, had the aggressive gait of a man used to getting his way. He had once thrived as a member of Saddam's inner circle. Like the man who would become dictator, he had been trained as an officer. As Saddam gained political power, al-Shahwani rose through the ranks of the professional armed forces, becoming a respected and brave commander, someone who would lead his men into danger rather than sit in comfort far from the fight.

However, al-Shahwani's exploits during the disastrous Iran-Iraq War put him in the crosshairs of the increasingly paranoid dictator, he later told his American friends, so he decided to leave the country in the 1990s for neighboring Jordan and then Virginia. In America, al-Shahwani bought a house in the suburb near CIA headquarters, and then he helped advise the Americans during the 1991 Gulf War. He remained on the U.S. payroll to head up a doomed attempt at a military coup against Saddam in 1996.

The regime's mukhabarat unearthed the plot when they caught a Sunni tribal sheikh carrying sophisticated communications gear into Iraq. Iraqi intelligence executed the entire network, including al-Shahwani's sons. In case there was any doubt that the network had been outed, the Iraqis sent the Americans a message using the equipment the mukhabarat had seized. Game Over, they wrote.

The collapse of that plot, however, did little to tarnish al-Shahwani's reputation among his American backers. They now had another reason to admire the Iraqi—his bravery and fortitude in the wake of his sons' deaths. Besides, during long afternoons barbecuing and drinking whiskey together, they had already concluded that al-Shahwani shared the same worldview that America's primary enemy in the Middle East was Iran, and its hardline Shiite clerical leaders must be fought by any means necessary.

By the end of the first year of America's occupation in Iraq, that view was what helped cement the CIA's decision to appoint al-Shahwani as his nation's new chief spy. The ideology dominant in Washington at the time was that America was facing a three-prong threat from an axis of evil, including Iran, Syria, and North Korea.

From the moment his appointment was announced in April 2004, al-Shahwani settled into the trappings of power. His hard-eyed bodyguards became notorious in Baghdad's government district, the Green Zone, for their arrogance and aggression. His convoy of armored Suburbans drove him to work each day at the Saddam-era mukhabarat headquarters, a yellow twelve-story concrete structure built in the same Soviet design found in Moscow's former client states around the world. The agency was among the most powerful and well-funded government departments in the country—for three years, American taxpayers paid $1 billion per year for everything from the new industrial carpeting to the metal-framed desks, ballpoint pens, and track lighting on the ceiling.

Before long, al-Shahwani's men started referring to their agency as Al Sharakat, or "the company," mimicking the nickname their counterparts at Langley had for their own shop. There was just one glaring problem with the setup.

Al-Shahwani had carte blanche to hire whomever he considered reliable and necessary to keep the nation safe. In an era of rapidly destabilizing security, that meant, in practice, that he gave many Saddam-era mukhabarat officers their jobs back. This was the very same cadre of men who had spent decades hunting, torturing, and arresting the people now running the country, who, like al-Shahwani, were also Sunni.

The new mukhabarat's focus on spying against alleged Iranian threats, as opposed to the Sunni jihadi terrorists killing innocent Shiites, poisoned al-Shahwani's relationships with the new Iraqi governing authority, which was dominated by Shiites. The Americans in Iraq seemed unaware or unconcerned with the implications of this situation. It was only after the chaos of insurgency engulfed the nation that the Americans realized they had the wrong man in the job.

Chapter 5
Ache for Paradise

For more than a millennium, the golden dome of Samarra's al-Askari Shrine glowed with sacred light over the fertile plains of Iraq's heartland. The architectural marvel was the pinnacle of human ingenuity, the largest dome erected in human history gracing the largest city in the Islamic world. Buried underneath the one-hundred-fifty-foot, intricately tiled structure were two relatives of the Prophet Muhammad, two of the twelve imams central to the Shiite faith, who died at the hands of the Sunni Abbasid Caliph during the wars of succession between the Sunnis and Shiites in the ninth century. According to the Shiite faith, the twelfth imam will bring salvation to the world when he emerges from a crypt underneath the shrine and leads the pious to paradise. To the faithful Shiites—whether farmers or poets, doctors or housewives—whose lives have been torn apart by war or trampled by oppressive dictators, the shrine had radiated solace as a reminder that their ache for paradise would be fulfilled.

When the Americans invaded Iraq, the importance of its sacred places was completely underestimated. But by early 2006, as Iraqi politicians squabbled for power in Baghdad and American officials struggled to bring order to a country riven with sectarian violence, Al Qaeda was preparing attacks that its leader in Iraq at the time, Ayman al-Zarqawi, hoped would reignite the blood-soaked theological battles that had raged in Iraq twelve hundred years earlier.

In a clandestine meeting held in January 2006, just a few miles outside Samarra and only a few dozen miles away from one of America's largest military bases, Al Qaeda's top commanders selected the sacred city as the site of an attack to trigger this new civil war.

In the chilly predawn air on February 22, eight militants snuck through the tight passageways of Samarra's old city, twisting through the empty lanes and sneaking past the small, family-owned shops locked up for the night. The men—two Iraqis, four Saudis, and two Tunisians—wore stolen Iraqi military uniforms, allowing them to move unchallenged. The city had been under a curfew for months and no one wandered the silent, narrow streets without permission. They reached their target as planned and swiftly overpowered the sleeping guards, leaving them alive but tied up and locked in a storage room. Then the terrorists laced a necklace of sophisticated explosives around the sanctuary, a display of expertise that would impress and horrify the American coalition's own ordnance teams. At 6:55 A.M. the dome exploded like a falling star, its seventy-two thousand golden tiles blown to pieces, leaving a yawning maw of wreckage and despair.

The force of the blast reverberated across the frosty plains. Samarra residents woke thinking an earthquake had struck. In a sense, they were right. As Samarra's golden dome shattered into millions of jagged pieces, so did Iraqi society.

Enraged at the destruction of their sanctuary, Shiite mobs rampaged in the streets of Baghdad and as far south as the port city of Basra. By the end of the day, the Interior Ministry reported that twenty-seven mosques had been attacked in Baghdad alone and several Sunni clerics murdered. In one instance, armed Shiites claiming to be policemen reportedly removed two dozen Sunni prisoners from a jail and killed them. As the days went by, American military patrols reported finding dozens of Sunni corpses with their kneecaps drilled.

By the end of the week, according to U.S. military estimates, one thousand Iraqis had been killed. The civil war had begun, and with it came the realization that America's attempts of the previous twenty-four months to rebuild a functioning intelligence infrastructure in Iraq had failed.

Abu Ali al-Basri was in his hometown in southern Iraq when he heard the news about the shrine attack. For months he had been steeped in the process of reconnecting with family and absorbing the suffering they had endured during his years of exile: the funerals of relatives arrested and disappeared by the regime; the poverty due to unemployment for men in the family because of their affiliation with him, an enemy of Saddam; the widows who fell ill and died alone, with no answers to their unresolved questions about what had happened to their loved ones. In Basra, he was hundreds of miles away from the unrest raging in the capital, but he had what he thought was a well-tuned ear to the social realities of Iraq. For people who had lost so much during the Saddam years, the country's religious sites had become a balm, and by 2006, when Iraq was teetering on the edge of chaos, despair and desperation had made religious identity even more vital. The fact that both the new

Iraqi leaders and the American coalition authority had left Iraq's holy sites vulnerable to attacks by Al Qaeda was an epic intelligence failure.

For civilians trying to cope with the constant fear of the next terror attack, Al Qaeda's strategy was overwhelmingly successful. Indeed, while Baghdad's new political classes were jockeying for power in post-Saddam Iraq, Al Qaeda leader Osama bin Laden and the leader of the group's Iraq branch, Abu Musab al-Zarqawi, were implementing an effective campaign to destroy the nation with intricately planned attacks.

A Jordanian by birth and an Al Qaeda veteran by 2003, al-Zarqawi had decided his destiny was to fulfill the group's theological quest to build a new Sunni Islamic empire, a replica of the medieval governing authority called a caliphate. Al Qaeda's desire was to purify society to re-create what they considered the religious utopia of the eighth century, the time when the Prophet Muhammad lived and Islam flourished. The place al-Zarqawi decided to enact his revolutionary dream was Iraq, and to make the plan work, he believed, his followers must kill Iraq's non-Sunni majority, the Shiites, as well as Christians and other minorities.

The logic was simple and brutal—the more Shiites and Westerners Al Qaeda killed, the quicker al-Zarqawi's dream would be realized.

The Jordanian terrorist attracted thousands of fanatical foreign Sunni Muslim fighters, but just as important, his campaign gained steam from cooperating with Iraq's Sunni insurgents who weren't religiously motivated but were filled with rage over the loss of power and prestige that they had enjoyed under Saddam Hussein.

Iraq's national security council meticulously recorded the obscenely large death tolls week after week: On December 19, 2004, twin bombings killed 70 in Najaf and Karbala, cities that are home to two of the world's most revered Shiite shrines. A car bomb exploded in Baghdad outside the headquarters of the Shiite SCIRI party, killing 13 people and narrowly missing a senior Iraqi Shiite political leader, Abdul Aziz al-Hakim. On February 22, 2005, during the Shiite religious celebration of Ashura, al-Zarqawi's operatives carried out five coordinated suicide attacks across Baghdad that killed 39 and wounded 150. A week later, a suicide car bomber dispatched from Anbar by al-Zarqawi's cousin killed 122 Iraqi army and police recruits in the Shiite-majority city of Hilla, the town where months earlier a football-field-size mass grave had been uncovered, revealing the bodies of those who had tried to overthrow Saddam at the behest of the Americans in 1991.

At the time, the Hilla attack was the single worst terror bombing of the war.

In weekly meetings designed to coordinate intelligence and security plans for the country, neither the Americans nor the Iraqis could map the networks of enemies destroying the fabric of the nation. U.S. forces in Iraq, meanwhile, didn't have the human resources or personal contacts to create a team of informants to track who was building and sending suicide bombs to the capital. And no Iraqi intelligence agency knew how to uncover or track an insurgent cell. Not even al-Shahwani, with his more than two decades of U.S. training and a $1 billion annual budget.

During these meetings, as Iraqi security officials sat with their American counterparts around a large oval walnut conference table in the prime minister's offices, many were reduced to gallows humor. Some would joke that Iraqis should consult with a fortune-teller rather than rely on weekly intelligence reports to describe the violence engulfing the nation.

Few societies have ever experienced the raw anarchy that followed the American invasion of Iraq—when police forces disappeared, bureaucrats who oversaw the simplest of government functions had no offices or managers to report to, and Americans raised in small towns in Ohio and California attempted to create provincial budgets for a country they had never been to and where they didn't speak the language.

In our imagination, war is all about battle plans and field combat, not the tedious management needed to build institutions or plan a reconstruction project. In Iraq, terror attacks and lawlessness had quickly become the dissonant rhythm of life, robbing the Americans and international officials in the country—who were also being targeted in the violence—of the space necessary to plan Iraq's redevelopment.

Meanwhile, wives and mothers of the one million Iraqis who had been disappeared under Saddam were trying to find out what had happened to their relatives. The former government operatives who had been too terrified to tell anyone where the mukhabarat had dumped the tortured remains of prisoners finally began to reveal those secrets.

With Saddam toppled, morning radio shows and the television evening news had regular features about mass graves that were being discovered weekly on farms, in abandoned buildings, or in dried-up lake beds. Travelers driving south on the highway from Baghdad could see hundreds of people digging in fallow fields, not for vegetables but for bodies. So many people had been seized and disappeared under the dictator that this grisly issue, as much as any other, captivated the nation. Shiite mosques across the country displayed lengthy, handwritten lists of names of people who had been taken by Saddam's secret police in case gravediggers came across their corpses. Widows pooled their money to rent a minibus on the strength of rumors of a new mass grave. No one knew exactly where their relatives had been killed. Mukhabarat records, such as they existed, were not made public.

It was clear to everyone standing at the multitude of checkpoints waiting to enter the Green Zone that the vast majority of al-Shahwani's employees at the mukhabarat, zipping past in their government-issued cars and with their VIP passes, were the same men—the same Sunnis—whom many Shiite politicians considered the architects of Saddam's torture squads. At a time when Saddam's old guard were claiming credit for setting off bomb attacks in Baghdad and Sunni sheikhs were writing checks funding Al Qaeda terror cells, it was no wonder that many Iraqis feared that Saddam's henchmen were trying to make a comeback.

Even before the Samarra shrine attack kicked off the civil war, senior American officials in Baghdad, including Paul Bremer, the head of the Coalition Provincial Authority, sent countless memos to the CIA asking for basic information about the motivation and identity of the men killing American soldiers. Yet the CIA station in Baghdad, the agency's largest outpost since Vietnam, sent back little to no information. One ambassador likened it to yelling into a dark well.

Instead of a unified effort to identify the networks of Sunni jihadi terrorists attacking U.S. soldiers, international aid workers, businessmen, and ordinary Iraqis, the American authorities running Iraq had buckled with the complexity of the task. U.S. military interrogators who captured Sunni jihadis would note their names and background information, but they had no centralized database to share their findings. That meant that some high-value targets, or even men like Abu Bakr al-Baghdadi—a minor cleric when arrested but who would later become the leader of the Islamic State—could be captured and released without anyone understanding their significance. Within the CIA, meanwhile, the bulk of the agency's staff in the country weren't there to tackle the network, which was rapidly becoming Al Qaeda's most deadly branch office. Instead, they were hunting the nonexistent stockpiles of weapons of mass destruction to justify the rationale that sent America to war in the first place.

In December 2005, Iraqis chose their first government. But six months later, in May 2006, with the violence after the Samarra attack threatening the fabric of the nation, the ruling coalition dumped its first post-Saddam prime minister, Ibrahim al-Jaafari, and rallied around a new leader, Nouri al-Maliki, the same man whom Abu Ali al-Basri had worked for at the Dawa party offices in Syria twenty years earlier.

A nakedly ambitious politician, al-Maliki knew he had only a short time to establish himself as an effective leader. In his first week in office, he ordered an emergency meeting of his national security staff and American military commanders. He demanded a progress report about Samarra.

It was clear to the Americans and to most Iraqis that the shrine attack had all the hallmarks of an Al Qaeda operation, given the expertise needed to assemble and place the explosives used in the bombing. By that time, the global terrorist organization had gained a reputation not only for its audacious acts like that of 9/11, but its sophisticated technical knowledge. What was missing when al-Maliki took over the reins of government in May were solid leads on the individual bombers themselves and the mastermind of the attack.

Everyone gathered at Prime Minister al-Maliki's oval table, regardless of partisan affiliation or sect, agreed with the assessment—with one notable exception: Mohammed al-Shahwani.

A hushed silence fell across the room as al-Shahwani's adjutant passed around a twenty-five-page report. The mukhabarat's conclusion was that the attack in Samarra was not the work of Sunni terrorists, but a false-flag operation conducted by Iranian agents with the express purpose of further discrediting Iraq's Sunni community.

Any scrap of credibility that al-Shahwani still had among members of the Iraqi government completely disappeared when they read that report. For most Shiite leaders, that document was the final proof they needed that al-Shahwani was incompetent at best, or, at worst, callously aiding and abetting the murder of Shiites by failing to recognize the clear and present danger of terrorism posed by Sunnis.

Al-Maliki left the meeting frustrated and enraged. No one had any answers, but more dispiriting was his feeling that he needed to stop the bloodshed, but that he didn't have anyone that he could trust to get him the intelligence necessary to do it.

A month later, American forces successfully found and killed the leader of Al Qaeda in Iraq, Abu Musab al-Zarqawi. But that didn't end the violence. The bloodshed only got worse over the summer.

In the first week of July, a car bomb struck the Az Zahra Mosque in west Baghdad, killing Shiite worshippers there. The next day, Shiite gunmen retaliated with a mass killing of fifty Sunni men, women, and children, plucking them from cars on the commercial street of the Rashid district, along the road to Baghdad International Airport. Neighborhood residents reported seeing the victims' bodies in the streets with their hands bound behind their backs, some with gunshot wounds to their heads, others with bodies pierced by bolts and nails. The city morgues reached their capacity after another one hundred fifty people, across Baghdad, died in revenge killings in a week and they stopped taking any more corpses.

The growing chaos made al-Maliki desperate enough to reassess his security team. He believed al-Shahwani was damaged goods, so the prime minister started assembling a team that he could rely on, men like Abu Ali al-Basri.

Chapter 6
Murder Capital of the World

When it was founded, the Iraqi capital had been the center of the civilized world, as magisterial and consequential as London or New York today. Beginning in the eighth century, walks along the crests and curves of the Tigris River led to debates about the meaning of life, the discovery of the numeral zero, the creation of the first astronomical observatory, and the aspiration to build the largest library the world had ever seen.

That was the Iraq Abu Ali had described to his young children as he tucked them into bed. That was the ideal that he had aspired to when he left them and Sarah in Sweden and returned to their homeland in 2003. But by late 2006, during his evening chats over Skype to his family, he had a hard time reconciling that dream of Baghdad and its reality. The city had become the murder capital of the world, transformed by sectarian militias and Al Qaeda, consumed by elemental urges of fury, vengeance, and lust for power.

Those were the pictures that Sarah and the children saw on their television screen each night. Sarah had a long list of reasons why she and the children would not return to Iraq. In the year and a half since the bombing of the al-Askari Shrine in Samarra, suicide bombings and attacks had risen by 74 percent. Nearly sixteen thousand unidentified bodies had stacked up at the city morgue, too mutilated for identification. A curfew kept Baghdad's four million residents cowering at home from sundown to sunrise. Sectarian infighting had paralyzed Iraq's first democratically elected government. Battalions of American soldiers patrolled city streets, turning themselves and Iraqis into targets for Al Qaeda.

Amid that hellish landscape, Abu Ali had one of the most difficult conversations ever with his wife in their twenty-seven-year marriage. Earlier that summer, his old boss, now prime minister, Nouri al-Maliki, had asked him to help restore order in Iraq's chaos. Sarah was focused on the bloodshed, images of the bloated corpses tangled in the reeds in the Tigris. Abu Ali, however, saw the hand of Providence. The job offered him a chance to make a difference.

And, as he kept telling his family on their long-distance phone calls, his new job as chief of security for the prime minister's office provided him with the most precious commodity of all in Iraq at the time: safety.

For the first time in his life, Abu Ali held a position of power, a far cry from the man who once hid in railway yards and lived hand-to-mouth. In his new job, he cruised through Baghdad's clogged streets in a government-issued armor-plated white Land Cruiser with a phalanx of bodyguards that whisked him past the hulking concrete walls that had grown around the city like kudzu, wild and unchecked. His driver flashed his precious blue pass at checkpoints, allowing them instant passage, instead of waiting in the snaking lines of teachers, civil servants, and nurses trying to get to work or buy their groceries and praying to get past the intrusive, humiliating searches before a suicide bomber could attack.

As his convoy, sleek and imperious, passed these lines of ordinary people, Abu Ali didn't feel the pride he had felt when he first accepted al-Maliki's job offer. He often imagined that the less fortunate Baghdadis that he drove by on his way to work cursed him under their breaths, the way his family once did Saddam's men.

The truth, Abu Ali thought, was that the people were right. The government had utterly failed in the basic task of keeping them safe. Al Qaeda terror cells carried out suicide bombings with impunity. Shiite death squads from militias supported by the son of the martyred ayatollah, Muqtada al-Sadr, roamed the capital in broad daylight killing Sunnis and cleansing whole neighborhoods of their sectarian enemies. Other Shiite militias had hit lists targeting prominent Sunnis for allegedly working for the previous regime. And Sunnis—with funding from tribes in west Iraq and sheikhs from Saudi Arabia and the Gulf states—were fighting back. Meanwhile, the much-loathed mukhabarat was back in business. Abu Ali's worst fear was that his family and the people on the street would associate his job with the agency whose mere name in Arabic conjured the sound of a baton breaking a man's ribs and the stench of a cell where human beings rotted away in agony.

What Sarah didn't know, and what few even in the hallowed streets of Baghdad's new government district knew, was that Prime Minister al-Maliki had tasked Abu Ali with a specialized responsibility. He wanted his old colleague to build an autonomous intelligence unit to counter what he and the international community viewed as the greatest threat—Al Qaeda.

It was a highly unusual arrangement, spurred by highly unusual times. The country was ablaze in violence, and the prime minister didn't trust his intelligence director, al-Shahwani, to analyze the dangers or provide him solutions. Al-Maliki couldn't fire the mukhabarat chief without a serious rupture in the U.S.-Iraqi relationship. Washington was paying tens of billions of dollars each year to rebuild the country and security forces. The price for that aid was keeping their favored men in positions of power. Al-Maliki, therefore, sought a way around him. Abu Ali was to build a team apart from all other institutions that would be answerable directly to al-Maliki.

During their phone calls Abu Ali explained to Sarah that he had a chance to right the wrongs of Saddam's era and prove that Iraqi intelligence could keep its citizens safe, instead of provoking fear. What he didn't disclose to her was the treachery he encountered almost every day within the national security apparatus, the campaign to undermine him by rival security commanders who were telling the Americans and the international community that he was working on behalf of the Iranians.

The vacuum of security unleashed a destructive energy across Baghdad, much like one of the notorious dust storms that sweep from the desert across the Tigris floodplain and descend like a biblical plague. Iraqis know that when the skies turn yellow it's a warning of the coming danger; they should run indoors to escape the torrent of pounding sand and dirt strong enough to strip paint from cars. Unlike the storms, however, the sectarian violence at the end of 2006 appeared unending. The American military decided on a defensive course of action: they turned Baghdad into a besieged garrison, erecting miles of twelve-foot-tall concrete barriers and fencing off whole neighborhoods from each other.

As a result, Abrar and the al-Kubaisis, like all residents of west Baghdad, found themselves cut off from most of the capital. Where for centuries the Tigris had created separate but diverse communities on the east and west sides of the city, concrete barricades reinforced ethnic and sectarian divides.

When Abrar commuted to her classes at Baghdad University, a journey that used to take fifteen minutes, she was now forced to endure a ninety-minute nightmare. Mountains of reinforced concrete surrounded government ministries like the walls of a medieval fortress. Roads navigable one day would be closed the next, on orders from a government bigwig afraid of being kidnapped or of being targeted by an insurgent's car bomb. Ali, the al-Kubaisis' neighbor, still drove Abrar and her father, who continued to lecture at the university. Abrar would peer out of the back-seat window and try to describe to her blind father how the entire geography of his hometown had shifted.

Professor al-Kubaisi tried to keep his apprehension to himself. But he didn't know where he fit in it anymore. His family, so smart and successful, once reflected the best and the brightest of the capital, the pinnacle of Arab success. Now, the al-Kubaisis were afraid to set foot in the Al Khassaki sweet shop, where they had always bought the city's famous nougat candies or, God forbid, enter the Ministry of Interior for new documents, because of the unmasked, almost feral, hostility that they, as Sunnis, would face.

Even on campus, Abrar could see how seriously the infection of sectarianism had taken hold. Students related to Shiite politicians and high-ranking members of their political parties were paying administrators to inflate their grades, a phenomenon she observed when exam results were posted at the end of each term. Abrar knew she would rank at the top. She spent more time than anyone in the chemistry labs, completing assignments in record time. She also knew that she had aced the exams themselves. There wasn't a single question that she didn't know the answer to. But as she stood in front of the bulletin board and read the exam lists, her name wasn't on top. Students she knew who didn't even understand the basics of the lesson plan had placed in the top 10 percent of the class, with the same marks she had been given. The top student was a Shiite whose father was an influential member of parliament.

Many of her father's students from Sunni families, especially those from the upper-class neighborhoods in western Baghdad like Yarmouk and Mansour, were leaving the country, driving to neighboring Jordan or even Dubai, places where Sunnis could hold their heads high, run their businesses, and live without fear.

But that wasn't an option for Professor al-Kubaisi. His disability meant he wouldn't easily find work in another country, and his wife's Shiite heritage might make them unwelcome in those countries. Starting over with no connections or relatives in a different city was a mountain too steep to contemplate.

In Sadr City, the al-Sudanis didn't have to fear the sectarian strife consuming the rest of Baghdad. Their sprawling slum had always been a Shiite ghetto. In fact, one of the most violent Shiite militias, the Sadrists Jaish al-Mahdi, considered Sadr City its headquarters and had erected its own defensive ring around the district. However, because the al-Sudanis weren't followers of the militia's spiritual leader, Muqtada al-Sadr, as many of their neighbors were, they were vulnerable to the whims of the movement's armed thugs.

By 2006, the battle of wills between Abu Harith and his oldest son was over. Harith had surrended and truce lines had been drawn. But in defeat, Harith had turned sullen. As part of the accommodation with his father for giving up on Nisreen, Harith acquiesced to the arranged marriage his parents desired. He married Raghad, a young woman with luminous pale skin and round brown eyes from a well-regarded Shiite tribe, and by 2006 he was already a father of two. Because of his dismal performance in university, Harith had few job prospects, but he needed to find a way to feed his own young family.

Abu Harith stepped in to solve that problem as well, calling in all the favors owed to him over the decades and spending this most precious social currency to get Harith a steady job as a bodyguard for one of the government's new ministers. It was not the kind of work that Harith had ever envisioned. He liked his colleagues but found that his job title belied the mundane reality of his daily tasks. Mostly, he sat in the vestibule of the minister's office, drinking tea and waving away petitioners, and in the evenings he spent hours with his team members at an exclusive government gym. The best part of the job was that the minister worked late and traveled frequently, so Harith had an excuse to spend long hours away from home, where his father loomed over him and the wife he never wanted waited with problems he couldn't solve.

Raghad didn't know what Harith had been like before she'd met him. The first time they met was at their engagement party, when she was eighteen. She had grown up in a small town in southern Iraq, near the village where Abu Harith was born. Their families were distant relations, and although hers had more illustrious tribal ties and stature in their home district, her father had been anxious to expand his social networks in the capital. Abu Harith had a good reputation, known as a rod of strength and moral probity, and for years everyone back home had heard his boasts about his clever oldest son, the one destined for a university degree and successful life. She didn't know that those plans had soured.

When she arrived in Baghdad to start her new life in the two-room apartment that Abu Harith had built for them on the third floor of the al-Sudani family home, Raghad discovered she was just the consolation prize for a man whose romantic dreams had been dashed. Her new life in Sadr City was a gray-hued purgatory.

Raghad rarely saw Harith and, in the brief periods when he was home, the floors hummed with tension, like she was walking over a live electric cable. She spent hours cooking his favorite meals, pressing his uniform to his exacting standards, and polishing his shoes. But nothing she could do made him comfortable. He spent the night on a foam mattress on the floor and only touched her when his parents' nagging for grandchildren became unbearable.

Raghad had done her duty as a wife, giving birth first to a girl, who resembled her side of the family, and then soon after a boy, who took after Harith's brothers. But even this wasn't enough to keep Harith home. And without him around the house, Raghad's own life shrunk further.

Growing up in rural Iraq she never considered the freedoms that she had taken for granted. Everyone in the village knew her and her family and no one questioned her motives for stretching her legs on walks through the fields. In Sadr City, the neighbors watched her every movement on the rare occasions when she left the house. What kind of woman would agree to marry a failure, the man whose Kurdish whore had driven a wedge in such a respectable family?

These whispers trailed Raghad each time she left the al-Sudani home. Each time a strand of hair slipped from her veil, every time her fingers brushed against a piece of fruit, she was scrutinized head to toe by neighbors whose appetite for scandal was never sated. When Harith didn't come home for days or weeks at a time, the murmurs grew. What has the woman done to drive him away from home?

Over time, as the other al-Sudani brothers married and had children, Raghad felt more alone than ever. Her brothers-in-law spent hours playing with their children, laughing with their wives. Her children noticed the vacuum in their own lives. At night they cried with longing. Why didn't Baba come home? Why didn't Baba bring them presents? Why didn't Baba play with them like their grandparents and uncles did? Raghad didn't know what to tell them.

While Harith's fortunes languished, Munaf al-Sudani soared. The younger al-Sudani considered 2006 a banner year, for that was when he received his acceptance letter to Iraq's newly reopened police college, a sprawling campus where thousands of Iraqis were training for positions within the security forces. There he learned ballistics on the shooting range and criminal detection techniques from retired American and Jordanian police officers and, on long evenings at the gym, built friendships as well as his physique.

His enthusiasm was contagious. Both he and his bunkmate had placed at the top of their training class two years in a row. They were part of the select few at the academy who had dreamed since they were boys of becoming police officers. Most of their fellow cadets were there less out of desire to bring law and order to Iraq than for a job that promised a steady salary and higher social status. The country's newly elected Shiite politicians viewed the new police forces as a patronage machine for their voters, especially the supporters of the prime minister, Nouri al-Maliki, and Muqtada al-Sadr, whose militia controlled Sadr City. In the quest for power, factions within the new Shiite elite considered the battle to control ministries, neighborhoods, and industry a blood sport. Iraqis had gotten rid of Saddam's corrupt bureaucrats, but a new crop was growing like weeds in an untended garden.

The more prosaic motives of his fellow cadets didn't dampen Munaf's desire to help his country, however. The nation was at war. The capital was being terrorized by suicide bombings and mortars falling on civilian neighborhoods. Iraq's new police force was necessary for a variety of irregular tasks, from counterterrorism raids to bomb disposal to intelligence work. To Munaf, this was exactly the type of challenge he had wanted all his life.

U.S. soldiers stood guard on every block, part of a force of one hundred sixty-five thousand throughout the country, along with about thirty thousand contractors and five thousand British soldiers. Munaf wanted to stand with them and help rebuild his nation.

By the end of 2006, residents of east and west Baghdad like the al-Sudanis and al-Kubaisis were so completely separated by ethnicity that they could have been living in different countries instead of only five miles apart on either side of the Tigris. Violence, it seemed, was the only thread that united the capital's families and neighborhoods.

That November, in the middle of the workday, militants from the Shiite Mahdi Army, in a convoy of minibuses, pulled up to the Ministry of Higher Education, where Abrar's uncle worked. Armed with automatic rifles and grenades, the militia stormed the building, forcing employees to gather on the lower floors. The militants locked the women in offices, and then separated the men into groups of Sunnis and Shiites. They marched more than a hundred Sunnis outside at gunpoint and into their waiting vehicles. Waving their guns and forcing traffic to a halt, they drove away through the middle of Karada, a prosperous neighborhood of mini-malls and cafes where intellectuals and actors take tea. The whole operation lasted fifteen minutes.

Abrar's uncle was lucky—he wasn't on campus at the time. But the whole al-Kubaisi family spent days fielding phone calls from families desperate for news about their kidnapped relatives. The government had issued no statement. There was no one they could call, no one they could trust, to rescue them.

Days later, masked men from a Sunni militia set up an ambush at the Ministry of Health, a government agency controlled by the party affiliated with Muqtada al-Sadr. For two hours, snipers from surrounding buildings fired into the ministry and mortars rained on the complex full of civilians seeking help with medical benefits and administrators overseeing the country's hospitals. The government called in elite Iraqi troops under the protection of American military helicopters, a show of force that caused the militants to flee.

The government imposed an indefinite curfew on the capital, banning all vehicles and pedestrians from the streets, and closed Baghdad International Airport as well as Basra's airport and port. The general staff put the military on high alert, beefed up checkpoints throughout the city, and established a cordon around Sadr City.

Abrar complained to her friends that she had trouble sleeping. She would leave the room when Prime Minister Nouri al-Maliki appeared on television. He kept promising to make the country stable again. But as far as Abrar was concerned, he and his Shiite coalition were the source of the problem.

Chapter 7
Education of a Radical

Tasneem was the closest person in the world to Abrar. The sisters had spent every night of their lives together and knew everything about each other. Abrar knew the sound of her big sister's breathing in the middle of the night and how much she despised the shape of her nose as if they were her own feelings. They shared clothes, a hairbrush, and a Koran. Other times, Abrar couldn't believe they were related. They fought like cats. Tasneem was the daughter who loved to bake and clean and play dress-up, while Abrar spent her afternoons poring over books trying to find out what made leaves green and rivers flow. In the summer of 2007, her older sister had done what Abrar always knew she would. She got engaged to a civil servant who hailed from their father's hometown of Hit, in Iraq's western province of Anbar. Tasneem was going to get married, have babies, and spend her days fussing in her kitchen and decorating the guest room—it was all she had ever wanted.

Everyone agreed it was a good match. The young man's family was related to the mayor and belonged to a strong tribe. Tasneem thought he was handsome. Abrar was ecstatic to learn that the newlyweds would remain in Baghdad, so that she wouldn't lose her sister entirely. God willing, she would be able to visit Tasneem on the weekends. Or whenever her husband allowed guests.

When the engagement was finalized the al-Kubaisi house turned upside down. Abrar could hardly study at night for all the activity and noise. Her sister, mother, and aunts spent hours discussing exactly what a new bride would need and how to spend Tasneem's dowry. One sat at the dining room table and sewed Tasneem's dress; another embroidered pillows for her wedding bed, giddy with anticipation. The wedding was going to be an elaborate affair, and dozens of relatives would travel from Hit to the capital.

In Iraq's Sunni heartland, people lived more easily than at any time since 2003. The surge of American troops through their provinces earlier that year had broken Al Qaeda's grip on the countryside. Iraqi tribesmen were getting paid by the Americans to protect themselves from the terrorist scourge. These jobs meant that families had money to put toward their sons' futures, to pay dowries and tuition fees.

Abrar had always thought her sister was the most beautiful creature in the world. On the day of her wedding, Tasneem looked like a movie star. Her plump cheeks glowed with just enough rouge, her dark brown hair was slicked back into a high bun, and her arms were decorated with the rich cedar brown designs of henna. The women of the family had daubed her décolleté with silver glitter to match her eyeshadow. Tasneem told her sister she spent the first weeks of her married life in complete bliss.

That year, one of the most important Islamic holy days, Eid al-Fitr, occurred in October, a few weeks after the wedding. Women prepare for days in advance, baking and cleaning for the three days of family visits and meals. On the morning of the first day of Eid, Tasneem called her parents to wish them a good holiday. She told them that she and her husband would be visiting her in-laws in Hit. She seemed happy with the prospect of the road trip. The following morning, the couple woke at dawn, packed their car with gifts and clothes for an overnight stay, and set off westward on the highway for the one-hundred-twenty-mile trip. They were expected to arrive before lunch.

They never showed up.

It took three full days for Abrar's father to piece together what had happened, days that Abrar can barely recall. The agony and shock of the news that Tasneem had gone missing was too much for any of them to bear.

The highways had been crowded that October morning, with thousands of travelers going home to visit relatives for the holiday. For the first time in years, the road the Americans had dubbed the highway of death was safe enough to contemplate such a journey. The U.S. military had given the highway that name because so many soldiers had been killed by improvised explosive devices (IEDs) while ferrying supplies to forward operating bases or while manning checkpoints that had been set up to try to keep Iraqi towns safe from Al Qaeda attacks.

These highway checkpoints, spread all across the country, soon became one of the deadliest places for civilians in Iraq. They knew from experience that the Americans, scared and trigger-happy, would shoot first and ask questions later. Between 2004 and 2010, the U.S. military tallied more than 14,000 incidents in which 680 Iraqi civilians had been mistakenly shot and killed and 2,218 wounded by U.S. gunfire at these checkpoints.

Professor al-Kubaisi thought that by 2007, after three years of occupation, the Americans should have known the social customs and the Islamic calendar. They should have known that the roads would be clogged with families. But for reasons the al-Kubaisis would never learn, on that early Eid morning, the U.S. Army had shut the highway leading to Ramadi, the largest town on the road to Hit. Hours went by and cars full of families, including Tasneem and her husband, were backed up for more than two miles, the road a jumble of vehicles. As usual, Iraqis considered lane lines to be mere traffic suggestions, and impatient drivers pushed into every available crevice, anxious to advance even one foot in the interminable delay. No one on the road was given any explanation for the closure. When pressure builds, the molecules need to escape. On that October morning, the emotions of angry and hungry Iraqis erupted into shouts and chaos.

Facing a surging mob, the Americans finally lifted the checkpoints. Frustrated Iraqis gunned their engines to push through the crowded road, Tasneem's husband among them. There was little room to maneuver around old pickup trucks and minibuses, but he was determined to make up as much ground as possible. It's unclear what exactly happened next. Some say the Americans opened fire. Some say a driver lost control of his vehicle. The end result, however, was a tragedy. Multiple cars flipped and Tasneem and her husband were crushed to death in a massive pileup.

When emergency workers pulled Tasneem's body from the wreckage, they surrendered her body to her husband's family, who, when the couple had not arrived by lunchtime, had traveled from Hit down the highway to see whether they had run into some sort of trouble. She was buried, per the Sunni tradition, before sundown that same night.

Abrar never got a chance to say goodbye.

Abrar didn't know how to move on. She felt a dull ache that flared at unpredictable times. A sweet, simple young woman had been killed, and no one seemed to care.

Her father spent a year struggling to get some sort of justice for his older daughter. He had his brother drive him to Ramadi, where the two men waited in a long line of Iraqis, all Sunnis, who were seeking compensation from the Americans. Each story that Abrar's father heard was more heartbreaking than the next. One man had lost all seven children when an American bomb hit their house. They weren't terrorists; they were all under fifteen years of age. Another man's brother had been slaughtered in the middle of the night because Al Qaeda considered him a traitor for joining a pro-American tribal militia.

When it was his turn, Professor al-Kubaisi walked toward the American officers, steadying himself with his cane in one hand and his brother on his other side. He had rehearsed what he would tell them, explaining his daughter's innocence, beauty, and the injustice of it all. It was impossible to put a value on her life, but he demanded the Americans help his family cope with their loss.

The military officers seemed unmoved by his speech. They told him to put his claim in writing, but not to get his hopes up. The U.S. military had no record of the accident, they told him, so they would not take responsibility. The chaos of the crowded highway that October morning meant that there was no way to know the identities of the other drivers on the road at the time of Tasneem's death. Since everyone in their car had perished, there were no witnesses, either.

Professor al-Kubaisi's kinsmen in Hit urged him to drop his crusade. They had lost a son, a nephew, and that was the real tragedy, they told him. In their way of thinking, male heirs were worth more than a daughter.

Others kept saying the accident must have been God's will. It was their duty as good Muslims to accept this and move on.

But Abrar couldn't accept it. She had just gotten used to her sister's absence from the house; now she was forced to live with her ghost. Tasneem's in-laws returned her trousseau. After all, what were they going to do with her clothes and linens? Tasneem's jewelry lay as it always had in the wooden box she had kept on the shelf in their shared bedroom. Her wedding dress was stuffed in the back of the small wardrobe. The embroidered pillows that their aunts had made were wrapped in plastic and stored under Abrar's bed.

The family was certain that Abrar would never get married. She was already twenty years old, already near the upper limit for eligible bachelors to consider, and she had always been intense about her studies. But after her sister's death, she had become more focused than ever on completing her degree. Her mother didn't want grief to consume Abrar. So, one afternoon she interrupted her younger daughter's reading with an offer she hoped might make Abrar happy. If she considered marriage, her mother told her, all of Tasneem's treasures would be hers. The thought appalled Abrar. In bed at night, the silence of the empty room felt like a weight on her chest. She missed her sister's quiet breathing. She even missed Tasneem's annoying habit of throwing her headscarves on the floor instead of folding them neatly, like Abrar did.

Lying there in the dark, Abrar couldn't sleep. Tasneem should have had a long, happy life surrounded by the pretty things that she had cared so much about. Instead, her life had been snatched away, and the people responsible for keeping her safe didn't care.

The long nights of insomnia took their toll. Abrar told her parents that one of her two brothers should take her room. They needed their own space, she said. She'd be comfortable enough sleeping on the living room sofa. She wasn't home much anyway. Her brothers didn't need encouragement: the younger one, Anis, took over her room the very next day. Abrar didn't think he knew that Tasneem's wedding pillows were under his bed. He never complained about ghosts. Maybe he just didn't care.

Abrar's new sleeping arrangement didn't help with her insomnia. But at least, in the living room, she could walk a couple of steps from the sofa to the desk and sit down at the family computer.

As soon as she turned on the clunky desktop computer and logged in online, the real world slipped away. She spent hours with people like herself, people outside of the mainstream, teetering on the edge and worried about the state of the world. She also researched scientific reports about cancer treatments. She clicked on a never-ending supply of videos showing dead Iraqis, victims of American military operations, and distraught Palestinian families whose homes had been destroyed by Israeli bombs. Online, Abrar read an Arabic-language copy of *Mein Kampf.* She started compiling lists of the greatest scientific minds who had come from the Arab world. In studying their groundbreaking work in their respective fields, Abrar noticed that they all had one thing in common. Like her, they were all Sunni.

Chapter 8
Building a Cover Story

Smoke twirled in fragrant tendrils upward toward the mildew-stained ceiling of the tea shop located in downtown Baghdad, just a few streets away from the country's new police academy. It was springtime, and the two brothers, Munaf and Harith, hunched over a low-slung wood table, each cradling a long-necked glass water pipe and drawing tobacco deep into their lungs for inspiration.

Munaf was twenty-three years old and had one more year left at the academy. He enjoyed the rigors of his new life, which required him to fulfill long hours of coursework and basic training six days a week while living in the police barracks. But earlier that morning, he had received an urgent phone call from their mother. Um Harith was scared. She needed Munaf to come home. There was just one problem: Munaf was a new recruit in Iraq's domestic intelligence agency, and it wasn't safe for him in Sadr City.

While studying to become an intelligence officer, the young man had learned how to battle Sunni jihadi terror cells, but from Munaf's point of view, the greatest threat to Baghdad and his family wasn't Al Qaeda. It was the men with whom he had grown up, the men who now filled the ranks of Sadr's militia, Jaish al-Mahdi, and who had turned his neighborhood into a combat zone. Since Saddam Hussein had been ousted, the militia had taken over his neighborhood like the Mafia. Armed young men roamed the streets at night, killing anyone they didn't like. They forced local businessmen to pay protection money and coerced local politicians into approving large government contracts with its members. The Jaish al-Mahdi reserved deep animosity for the federal security forces, people who worked for the Americans, and ordinary residents of Sadr City like the al-Sudanis who didn't belong to their political party.

By April 2008, Sadr City, with two million residents in an area half the size of Manhattan, had been completely walled off from the rest of Baghdad, with American forces on the perimeter trying to keep militia members from smuggling weaponry inside. Militia members, meanwhile, watched every vehicle entering their domain, paranoid about American spies.

The Sadrist threat was consuming more and more of the government's time. The militia had been killing American soldiers for years, but only recently had Prime Minister al-Maliki finally decided that the group's siege of Sadr City and its stranglehold on the nation's port city of Basra, from which most of Iraq's oil exports flowed, made it a national security threat.

Harith hadn't been home much in the previous year. His boss, the Iraqi minister for state security, had become one of the prime minister's trusted loyalists assigned to tackle the Sadr threat in Basra. Whenever he happened to be in Baghdad, Harith met with Munaf to chat and smoke. But Harith had no desire to go back to Sadr City to his loveless marriage, nagging father, and no peace.

Munaf felt differently. A few months earlier, he had gotten married. Years earlier, his mother had arranged a match between him and the daughter of a neighbor, one of her best friends. Munaf hadn't needed any coaxing. Nismah was one of the neighborhood beauties and the two of them had played together when they were children. Their wedding was small, by Iraqi standards, but that didn't matter to him. Munaf was enchanted with his new wife. He tried to get home every weekend, on his day off, to spend as much time as possible with her in the room his father had built for them on the third floor of the al-Sudani family home.

That spring, it had become harder and harder to make it back to Sadr City. Every evening, when Munaf called home from the academy, just four miles from his family's home, it felt like he was calling another country. He could hear the heavy boom of artillery as Nismah told him the latest frightening stories from the neighborhood.

The Jaish al-Mahdi had just executed a shopkeeper who had voted for Prime Minister al-Maliki's Dawa party, one among the hundreds of Sadr's Shiite political opponents killed in Sadr City during the past year. Rockets had landed at Jamila market, burning the whole place to the ground and causing food shortages.

Tensions escalated further, sparking the phone call from Um Harith and Nismah that morning. The militia had set up a mortar position down the street near Munaf's uncle's house, sparking a furious battle with the American soldiers.

No one could sleep, Nismah told him. We thought we would all die. Your uncle's house has been destroyed, all the windows smashed.

His mother tried not to sob when she took her turn on the phone. Son, it is terrible, just terrible. Whoever thought our people would be killing each other like this?

The women never asked Munaf to come home. They didn't need to state the obvious. Deep inside Sadr City's dense web of alleyways and lanes, his elderly parents and younger siblings, not to mention his and his brother's wives and children, were behind enemy lines. One of the eldest sons should be there as a bulwark against the armed thugs.

By rights, Harith should be the one responsible for the family's safety. But speaking with Harith in the teahouse, Munaf knew that it wasn't going to happen and that he would have to step into the breach. The question remained, however, how Munaf could get home safely, past the Sadrist checkpoints, without being held or killed by the militia as a possible government spy.

You'll need a disguise, Harith told him. The question is how can you disguise who you are?

It wasn't an easy question to answer. One look at Munaf and everything about him identified him as police. His academic success at the academy had enhanced his confident stride. Hard work at the gym had given him a chiseled physique. His cropped haircut, square jaw, and mirrored sunglasses gave him a striking resemblance to Erik Estrada, the American actor who played a cop on television.

Munaf knew it wouldn't be enough for him to wear different clothes and try to blend into the crowds at the checkpoints. If a member of the militia even suspected Munaf of being with the Iraqi security forces, they would shoot him, no questions asked. He needed props, something that he could use to create a credible covert identity.

That's when Harith had an idea. Drive my minibus, he told Munaf. Pretend to be a bus driver and no one will give you a second glance.

Harith's salary working for the Ministry of State Security was comfortable, more than his wife needed for their children. In fact, he also shared a portion of his salary with his parents to help offset expenses for his younger siblings. And he had followed the lesson he had learned as a teenager working with his cousin in Jamila market—the value of investing. Baghdad was a city of five million people, but the city's transportation demands far outstripped the abilities of the public bus system. Now, with the ever-changing landscape of checkpoints, blocked roads, and curfews, regular bus routes were constantly interrupted. Entrepreneurs like Harith had stepped in with a profitable solution—a network of minibuses that traveled between downtown and larger residential districts, a service that commuters called Kia, after the inexpensive South Korean–made vehicles they rode in.

Harith had been renting his vehicle out to a friend of the family who drove a commuter route each day. Sucking on his water pipe, Harith told Munaf that he would sideline the business so that Munaf could get home and check on the family on weekends.

Take the van, Harith kept telling Munaf. Just keep it and keep yourself safe. But he didn't ask Munaf to make sure their father understood that it was Harith who had figured out this scheme, and that it was yet another way he was doing his part for the family.

The following afternoon, a Thursday, Munaf prepared to meet Harith's friend who drove the Kia van.

He stood in front of the mirror and took a hard look at his reflection. His jaw was covered with a day-old beard. His long-sleeve shirt was thin from years of washing and stained with patches of motor oil. His fingernails were tattooed with dirt. The reflection staring back at him looked like a typical Baghdadi bus driver, nothing at all like the person Munaf was trying to become. If only my instructors at spy school could see me now, he thought.

The driver was already waiting when Munaf arrived

at the shisha cafe near the central bus station, bringing with him the minibus keys and a set of his own clothes—the al-Sudani brothers thought it would be simpler and more effective for Munaf to wear what an authentic driver would. As his trainers at the academy had always told him—the details of a cover story matter.

The driver smoked a water pipe with Munaf while he explained his job, telling him the intersections and traffic circles where he collected passengers traveling from downtown back to Sadr City, and how Munaf needed to behave at the checkpoint. The Americans were being shot at constantly, so unless you followed their directions exactly, they might shoot you first.

Passing the checkpoint would be easy. His ID showed his permanent residence in Sadr City, and since he was a local, the Americans would let him pass. The only thing that could go wrong was if a Jaish al-Mahdi militant pulled him over and demanded to know more about him. Munaf felt confident there would be no reason for anyone to do that. He appeared to be exactly what he was pretending to be—a bus driver and nothing more.

He shook hands with Harith's friend and set off to find his riders, something easily accomplished by stopping at the nearby Finance Ministry. With a few people in the back seats, it was time to test his cover and drive home.

Munaf had spent so much time preparing himself for any possible contingency that he was almost disappointed at the ease in which he traveled through the first checkpoint leading into Sadr City.

The American soldiers stopped the vehicle and a guard with a bomb-sniffing dog walked around it, scanning it for any explosive residue. The soldiers then demanded that he open the back so they could ascertain what cargo he was carrying, but they barely glanced at him. The same thing occurred at the Sadrist checkpoint. Harith had been right. No one pays much attention to a bus driver.

Munaf spent two nights at home with his family, his presence a balm for their stress and worry, and promised he would return again the following weekend. As the summer went on, Munaf started enjoying his personal undercover operation. He liked the playacting but, more important, during the journeys back and forth between his besieged neighborhood and downtown, he was also gaining a useful talent: acquiring and analyzing reliable intelligence.

Sitting in the long line of cars waiting to enter Sadr City on Thursday evenings, Munaf had a front row seat to some of the best intelligence that a security officer could ever hope for in a conflict zone. His passengers were mostly civil servants, returning home from their ministries, primarily men who pushed paper every day, or sometimes women who taught at one of the downtown universities. But they were also neighbors to some of the most violent men in the country, making them privy to the daily movements and mentality of the enemy. Small talk in the back seats of the Kia van mostly shied away from politics, focusing instead on information necessary to stay alive that weekend—whether certain streets had been cleared of militants, where Jaish al-Mahdi stored weapons caches, who in their neighborhood was planting the improvised explosive devices, and which of the local commanders had been killed.

By listening in, Munaf learned how to distinguish between mere rumors and reliable intelligence. Given the high stakes Munaf and his passengers faced when entering Sadr City, it was vital to know the difference.

By the time he finally reached home on those Thursday evenings, Munaf had acquired a gold mine of knowledge. But he had no one to tell it to. As a cadet, he didn't have a commanding officer. He didn't even have a formal role in the security forces. But he couldn't just sit and let his information go stale.

So he did the only thing he could think of. He called Harith. He always started these conversations the same way, offering his brother the respect of a younger sibling. He'd tell Harith that he had made it safely home, that his bus was safely parked, and that his wife and children were okay. Then he'd turn professional. He'd repeat what he had picked up about Sadr's militia. He trusted that Harith would pass along the intelligence to his own commander. After all, his brother was participating in the operation against the Jaish al-Mahdi in the south. Maybe someone from his minister's squad could put the Sadr City information to use. Maybe it could even save a life.

After the call to Harith, Munaf settled into his family's weekend routine. On Fridays, while his mother and his new wife cooked lunch, his father invited him to the majlis to sit next to him as he entertained the relatives who were staple guests each week. As the men ate, his uncles and cousins and younger brothers took turns discussing work or problems with their neighbors. Abu Harith, who normally gave advice on all such matters, would often turn to Munaf to ask his opinion.

It had happened in a blink of an eye. In Harith's absence, Munaf had transformed from the invisible son into the lodestar of the family's life. He had taken on the responsibility of ensuring their safety, and, in accepting that challenge, he had received his father's respect in return.

No one commented on this outright, but the interactions with his family made it clear that his standing had changed. Munaf sat across the sofra from his father during meals, the place reserved for his equals. His younger sister or even Harith's wife, when they served tea after dinner, made sure Munaf's glass was poured first, or at the same time as Abu Harith's. When the children fought over what channel to watch on the family's only television, they brought the remote to Munaf to arbitrate. When Um Harith needed more money for food, she came to him.

Harith's photos remained prominently displayed on the living room wall, but, for the first time in his life, Munaf felt like he was seen.

By the end of the summer, Munaf's weekend routine had become widely discussed around the academy. In the evenings at the gym, many of his fellow cadets called him crazy for traveling back and forth to Sadr City. Why was he taking such risks every week? Did he want to die before he even got his lieutenant stripes? Munaf joked that the trips would earn him a Ph.D. in street smarts.

The more he thought about it, the more Munaf realized how much he liked his improvised fieldwork, the thrill of holding his nerve and gathering information. When the new term started at the academy, he told his instructors that he wanted to join the Iraqi equivalent of the FBI, the domestic intelligence agency that reported to the minister of interior, rather than the CIA-funded mukhabarat.

Walking between lectures at the academy one day that fall, Munaf saw the skies turning a sickly yellow. A haboob, one of Baghdad's notorious sandstorms, struck with unusual ferocity. Outside his barracks, Munaf watched towering palm trees ripped from the ground as tightly packed balls of dirt and mud pelted the courtyard. For three days the storm raged, closing the airport and keeping everyone indoors.

As soon as he was able, Munaf went to retrieve Harith's minibus from its parking spot near the academy. The storm damage was immense. Roofs had been torn off street kiosks. Trees blocked roads. And the Kia had been crushed under a two-story metal advertising billboard, its windows smashed and the roof bent like a sardine can. He called Harith.

Munaf tried to soften the bad news with laughter. Brother. May God bless you and keep you safe, Munaf told him. Unlike how he has treated the Kia.

Motherfucker. What are you talking about? Harith asked him, his voice rising.

God gave us terrible weather this week. The haboob crushed the Kia. There's no hope for it, Munaf told him.

What do you mean crushed? Why are you bringing God into this? Munaf, what have you done to my car?

Harith, there is no need to get excited over that piece of junk. Believe me, you wouldn't miss it if you ever had to drive it. The car was a piece of shit.

The joke turned into a full-blown argument. Harith was furious that Munaf had left the car exposed. Munaf then accused his brother of doing the same with their family. Finally, Harith snapped. He didn't want to be scolded, especially not by his younger brother, about family obligations.

You want to be the responsible one? Harith kept asking Munaf. Well, then, buy your own car. Or pay me for the one that you destroyed.

Munaf hung up the phone, stunned by his brother's anger. He dialed his father next, to explain why he couldn't make it home that weekend. He was in for another surprise when Abu Harith took his side, saying he wasn't obliged to pay his elder brother for the damaged car. A responsible man pays his debts, Abu Harith told him. But all summer long you have been fulfilling your brother's obligations. I don't see any debt here.

A few years earlier, Munaf would never have expected his father to side with anyone in the family over Harith. The family hierarchy had truly changed for good.

Chapter 9
Learning from Misinformation

Abu Ali's office in the prime ministry complex was notoriously unkempt. Files and books were always piled in uneven stacks perilously close to toppling over and burying him at his desk. His subordinates were too afraid of their boss to file his paperwork. They knew that the chaos made sense to Abu Ali. Like a magician, he could pull the right form or report from his unruly stacks in an instant. One dog-eared file, though, was always easily found—the one for his pet project, the Falcons.

Inside was a hodgepodge of handwritten notes, some on computer paper, some in lined schoolbooks, with Abu Ali's research. The advantage of being an inconspicuous man amid loudmouthed political climbers was unfettered access to information. Abu Ali wanted men around him who agreed with his philosophy of intelligence-gathering as a science and not a blood sport. In his deliberate and methodical way, he had started files on people he might possibly recruit for his unit, young men who were performing at the top of their classes at the military and police academies and older men whose names he had overheard from commanders when they discussed operations.

By the summer of 2009, with his research file bursting, Abu Ali began contemplating ways to evaluate these potential recruits himself.

Abu Ali dusted off his tradecraft acquired from his years living on the run in Baghdad and went undercover to the two training academies, gaining access by pretending to be a workman—someone anonymous whom no one would notice or remember, but who would have a legitimate reason to visit these high-security locations.

It was easy, given his resources and power at the prime minister's office, for Abu Ali to pull off a convincing reconnaissance mission. Once, he falsified a Ministry of Electricity identification badge and went to the police academy posing as a technician looking to inspect wiring. Another time, he borrowed a truck from the Ministry of the Interior so he could pretend to be an inspector looking for invented structural flaws in the building.

One day in late June, Abu Ali went to the police academy disguised as a trash collector, a visit memorable because it was when he first heard about Munaf al-Sudani.

Who pays attention to a trashman? Almost no one, Abu Ali learned, as he wandered the hallways completely undisturbed. He could sweep up documents discarded by instructors and department heads without anyone looking twice at him. That late afternoon he walked by two officers taking a cigarette break and laughing with admiration about Munaf al-Sudani's forays into Sadr City as a minibus driver.

That boy has daring, the older man told his colleague. He's not easy to scare.

Abu Ali's ears perked up. When he got back to his office, he wrote the name down in his Falcons file. When he graduates from the academy, he thought to himself, this is a man for me.

In the meantime, urgent political and security matters pulled Abu Ali al-Basri away from his recruitment drive.

In the two years since Abu Ali had taken the job as Prime Minister Nouri al-Maliki's head of security, he had garnered a reputation as a man with eyes that missed nothing. When politicians and security commanders sat in the prime minister's office, he silently observed the complaints and theatrics. It was his boss's job to keep political leaders happy; it was his job to keep the city safer. When Iraq's dignitaries finally shuffled out to their next meeting, Abu Ali would return to the quiet of his own office to ponder which parts of the information he had just heard represented real security threats and which was posturing.

Yet that summer, the long hallways of the prime ministry complex had turned into a treacherous jungle overrun with political backstabbers and careerists. Everyone in Iraq had been clamoring for an end to the killing, from al-Maliki's party to his government coalition partners and the international community that was spending billions of dollars a year to rebuild Iraq. The prime minister, however, didn't trust many of the commanders from the nation's official security agencies. So he decided to form his own units, using money from his own discretionary government budget so he could bypass parliament. Al-Maliki's political supporters said his actions were legal; his detractors said the opposite. The prime minister had a special budget that was unaccountable to any oversight and shielded these special security units' actions from review as well. One of those units was Abu Ali's own, but there were several others, the most controversial of which was called Force 54. It was Force 54 that the Iraqi parliamentarians had come to complain about that day.

Time and again, Abu Ali had sat on the overstuffed sofas in al-Maliki's personal office while the prime minister received briefings from Force 54, his Praetorian Guard, also known as the Baghdad Brigade. The commanders had become a mainstay in the government complex as well as in Baghdad's security operations command headquarters. If the prime minister wanted someone arrested, whether a terror suspect or a political enemy, chances were that Force 54 took on the mission. They didn't usually bother informing any of the other services, however, or coordinating their actions. And that wasn't the limit of the unit's work. The brigade ran its own detention center in a corner of the old Muthana Airport, a short drive away from the Green Zone and the prime ministry building. Regardless of what his supporters claimed, the unit and its work contravened all of Iraq's new laws, the ones designed to prevent the crimes of the past, when security forces acted with impunity and tens of thousands of people were disappeared—like Abu Ali al-Basri's father, his grandparents, and his sisters-in-law had been.

Force 54's briefings for the prime minister were routinely upbeat, detailing how many terror suspects they had arrested and how much headway they were making uncovering leading Al Qaeda members. But that didn't make them true. Twice in two years the brigade had told al-Maliki that they had captured the head of Al Qaeda in Iraq, Abu Omar al-Baghdadi (not to be confused with the later leader, Abu Bakr al-Baghdadi). The prime minster's office gleefully publicized each of these supposed successes, the latest instance in April 2009, only to be embarrassed later when the news turned out not to be true.

The group's checkered record and reputation for using torture to extract information underscored the deep flaws in Iraq's intelligence agencies. Still, al-Maliki saw his official national intelligence service as compromised due to the large number of Saddam-era mukhabarat officers who had been hired by Mohammed al-Shahwani. He couldn't close down the national intelligence agency outright—doing so would be perceived as a direct attack against the American government that funded it. Instead, al-Maliki made it irrelevant. He allowed an arrest warrant to be issued against al-Shahwani, prompting him to leave Iraq for neighboring Jordan. He then flooded the national intelligence agency with thousands of new recruits—men who had no experience in counterterrorism work but who were members of his own political party. By 2009 the quality of the work coming out of the CIA-backed service was so unreliable that it was effectively sidelined.

While al-Maliki played politics, Abu Ali sought to deal with security. And Force 54 wasn't helping. However, the accusations by lawmakers from al-Maliki's ruling coalition, that his Force 54 were torturing fellow Iraqis, just as Saddam's mukhabarat had, fell on deaf ears. More than four hundred men were allegedly being held in Muthana jail, without arrest warrants or any legal oversight. The prime minister barely batted an eye at the news—he dismissed the accusations as smears by his political opponents.

Abu Ali al-Basri knew that the parliamentarians' accusations were true. Iraqis from Sunni areas were being rounded up and tortured by commanders seeking information on Al Qaeda cells. Their actions were as troubling to him as the fact that those commanders failed to see that their tactics were not making the country safer. But without political cover from al-Maliki, Abu Ali could not attempt to curb these abuses, let alone suggest punishment for those responsible. Instead, Abu Ali hoped his rival commanders would dig their own graves with mistakes.

In the meantime, he had urgent work to do. While the American military had broken the Al Qaeda insurgency across the north and west of Iraq, Baghdad was still being rocked by bombings. Al-Maliki hoped to win a new term in elections scheduled for January 2010 and he couldn't do it if the capital was under siege. He was desperate for splashy, positive news, like successful counterterrorism operations. He didn't want scandals like those that rival parliamentarians had brought him about Force 54.

That's why the prime minister held Abu Ali back when the meeting finally ended. In his usual terse style, he ordered his new spymaster to deliver something that al-Maliki could spin as a victory, instead of the false information from Force 54.

Abu Ali was incredulous. He had only three men under his command, and he had only six months to deliver his boss a high-profile terrorist. What al-Maliki didn't care to understand was the time and diligence required to assemble good intelligence, find reliable sources, and capture or kill the targets doing the most harm. Abu Ali wanted to prove once and for all that good investigative skills were what Iraq needed in the fight against terror, not fear, torture, or chains. He wasn't sure he had enough officers to fulfill this mission—but he did know whom he would hunt.

The Americans had decimated whole networks of insurgents in the northern Sunni provinces and western region with their surge of forces and alliance with Iraqi Sunni tribal militias. In 2007, U.S. forces had even tracked and killed the Iraqi Al Qaeda operative who had planned and spearheaded the horrific bombing at the Samarra shrine.

One of the most appalling terror attacks since the U.S. toppled Saddam Hussein, however, had not yet been resolved. On a muggy August afternoon in 2003, a suicide bomber rammed a cement mixer full of explosives into the wall surrounding the United Nations headquarters in the Canal Hotel. The volume of explosives caused the three-story hotel to implode, trapping hundreds of humanitarian workers under walls of concrete. Sérgio Vieira de Mello, the special representative of the U.N. secretary general, was buried under several tons of rubble, his legs crushed by debris. He stayed alive for several hours after the explosion, but rescue workers couldn't reach him. He was one of seventeen people who died in the attack, the first time that a humanitarian organization had been directly targeted in a war zone. The strike violated the rules of international warfare, under which international organizations and aid workers in a war zone have been treated as noncombatants, and led to the total withdrawal of these organizations from Iraq.

The sophistication of the attack and size of the bomb, like the Samarra attack, left no doubt that Al Qaeda had been responsible, but more than six years after the bombing, the mastermind of the Canal Hotel attack had not been found.

At the same time, the prime minister was receiving on an almost daily basis a laundry list of complaints from the international community about how his government was failing to meet the expectations necessary to keep their billions of dollars in financial aid flowing to Iraq. The prime minister was being blamed for the delay in school construction, the lack of progress on judicial reform, burgeoning corruption, and, of course, the dire security situation. Everyone would benefit, Abu Ali al-Basri thought, if the prime minister's office could show the United Nations a breakthrough in the 2003 case.

In June 2009, Abu Ali gathered his three men in his cramped office. They had spent six years in the new Iraqi security forces, but their ties to Abu Ali went back to the days of exile. They were smart, they were trustworthy, and, for the time being, they were all the resources Abu Ali had.

He told his men not to think about their task as a solution to the war against terrorism. For all the books, policy memos, and military field manuals written in Washington, London, and elsewhere about Sunni terrorism, counterinsurgency strategy, and intelligence-gathering, no one in Washington or London or anywhere else in the world had a formula for accomplishing that. Instead, Abu Ali told his men, they should visualize their mission like a falcon hunting his prey.

The first task was to identify a suspect. To do that, Abu Ali ordered his team to gather any leads Iraq's other security forces had on the Canal Hotel bomb, and to find out if any other agency had an active investigation underway, or suspects in custody. For years, snippets of information had been collected by different Iraqi services, among them the counterterrorism forces, the domestic intelligence agency, and Force 54, but none of the agency directors talked to each other—in fact, they despised each other. With bombs going off almost every day in the capital, the Canal Hotel attack was a distant memory in an era of unspeakable violence.

With the authority of the prime minister behind him, Abu Ali cut through the interagency deadlock. Within a few days, he and his men had a jigsaw puzzle of clues and, just as important, a name. An Iraqi prisoner convicted of being an Al Qaeda cell member in Baghdad had fingered as the prime suspect in the bombing a retired Iraqi commercial pilot, someone who had lived among Saddam's elite, but whose fortunes had plummeted when the dictator had fallen.

Abu Ali ordered a team from Force 54 to find the retired pilot. Not touch him. Not arrest him. Just find him. Once he knew where the man lived, then Abu Ali and his men would decide how to proceed.

For a man who allegedly had so much blood on his hands, Ali al-Zawi was living quite comfortably in a three-story home in one of west Baghdad's upscale Sunni neighborhoods, Yarmouk, a verdant district built for Saddam's military officers and families. Iraqi and American prisons were filled with Sunni terrorist suspects, but for years no one had thought about the Canal bombing—until Abu Ali and his men swooped down like the bird of prey for which they had named their team.

The spy chief knew that with so few men, he would have to work out a simple plan. The first move was to arrest the suspect and then determine whether al-Zawi was indeed an Al Qaeda cell leader, or whether the allegation against him was a case of mistaken identity, as it is a common name. The easiest way to accomplish this, Abu Ali thought, was to undertake a sting operation and get his men inside the suspect's home, to see what evidence they could find.

After a few days of surveillance, his team reported that their suspect had put his home up for sale. Abu Ali didn't want to take a chance that he might flee Baghdad, or worse, leave the country. So on June 26, he gave the green light for the operation to begin.

The plan was for two of his Falcons to pose as real estate speculators. As a prop to make their cover convincing, they borrowed $10,000 from the petty cash kept in the prime minister's office and went to Yarmouk aiming to get inside the suspect's home.

The two undercover officers, carrying the briefcase of cash, rang al-Zawi's front door. After a brief explanation, he invited them inside. Abu Ali was monitoring the situation from a car a couple of blocks away. The Falcons were there to start a conversation, using the real estate deal as a pretext, to get al-Zawi's guard down and to see whether there was any indication that the man was a jihadi sympathizer. They weren't supposed to escalate matters, but, just in case, Abu Ali had organized a backup team from Force 54, because if anything went wrong, Abu Ali didn't have the men necessary to protect his own.

It wasn't long before Abu Ali's men came back out to the street, but this time without the briefcase. Through his binoculars, Abu Ali saw the pair give the prearranged hand signal that al-Zawi should be arrested.

Abu Ali jumped out of the car and ordered the Force 54 team to the house. Their noise brought al-Zawi to the front door. As they flooded through his courtyard, the suspect didn't fight his arrest and left quietly with the guards.

Under Abu Ali's supervised interrogations, Ali al-Zawi stayed silent for weeks. Other than the one allegation made by the other prisoner about al-Zawi's role in the 2003 terror attack, the Falcons had nothing to confirm his involvement in the U.N. bombing. But the unit kept tabs on al-Zawi's house. Over the next six months they watched Al Qaeda couriers, cell leaders, and financiers use the residence as a safe house, dropping weapons and cash for their terror plots. Abu Ali finally ordered that the walls of the home be torn down, revealing hundreds of thousands of dollars in cash and materials for bomb-making. Faced with the growing evidence against him, al-Zawi finally confessed to planning the U.N. attack, and to organizing a separate bombing in Sadr City in 2008.

Abu Ali forwarded the names and locations of a half dozen other Al Qaeda suspects, men al-Zawi said were involved in his cell, to the Baghdad Operations Command. Prime Minister al-Maliki, meanwhile, had gotten the counterterrorism success he had been hoping for ahead of the national election. There was just one problem. When al-Maliki ordered state news outlets to publish the arrest, the United Nations secretary general's office didn't believe the news.

By the end of 2009, mistrust had corroded relations between top U.N. officials and al-Maliki and other diplomats in Iraq, because foreign officials knew the prime minister was politicizing intelligence ahead of the election. Furthermore, classified cables to London, Washington, and New York had concluded that al-Maliki was purposely stoking sectarianism as an election campaign tool and the leader was ignoring warnings of human rights abuses by Iraqi security forces. So when news of the al-Zawi case became public, there was widespread skepticism among many foreign diplomats that the Iraqis had gotten the right man, especially as al-Maliki refused to share Abu Ali's intelligence methods.

The news, however, made the Americans take notice. If there was an Iraqi professional tracking down high-level Al Qaeda members and sympathizers, the

U.S. special forces doing the same thing wanted to know who they were.

From al-Maliki's perspective, Abu Ali al-Basri had delivered the positive security news he had craved, and after the al-Zawi case, the spymaster received his coveted mandate to recruit more members into what he had started referring to as the Falcon Intelligence Cell. As 2009 drew to a close, he dug back into his dog-eared file of candidates and started making contact.

On a crisp early-November morning, some five hundred Iraqis, including Munaf al-Sudani, stood in formation on the police academy parade ground for their formal graduation ceremony. They saluted while a band played Iraq's new national anthem, and remained at attention while government luminaries addressed them from a raised, covered podium.

Abu Ali sat in the second row, with his official bodyguard escort in attendance, as the official representative of the prime minister's office. One of his guards was down on the dirt-packed field waiting for his boss's order.

When the speeches were over, the cadets let out a cheer and threw their berets in the air. Abu Ali signaled to his guard, who weaved through the young men, looking for Munaf. He found the lieutenant surrounded by friends. His dark blue uniform identified him as belonging to the domestic intelligence agency, known in Arabic as Istihbarat. His cropped, spiky black hair and muscular body were supposed to make him look like a tough guy, Abu Ali's guard thought. But his wide smile ruined the whole effect, making him look like a little kid.

Congratulations, the bodyguard told him, as he tapped Munaf on the shoulder to get his attention. My commander would like a word with you. Could you please follow me? He led the new graduate back through the crowds to Abu Ali's gleaming Land Cruiser.

If Munaf was nervous as he stepped up into the cool leather interior of the car where Abu Ali was waiting, he didn't show it. His equanimity made Abu Ali like him even more. Salaam wa Alaikum, he said to the new lieutenant. How would you like to come to work with me?

The proud spymaster told Munaf about his mission. We're saving lives, Abu Ali told him. And we are doing it in a way that is professional and preserves our integrity. He told Munaf general details about the still classified al-Zawi sting operation as the kind of work he would be involved in if he joined the Falcons. The pitch convinced Munaf, and eight other new recruits as well.

By the time of the January elections, the Falcons had grown to a team of thirteen, Abu Ali and twelve dedicated men.

As Munaf al-Sudani was settling into his new work with the Falcons, across Baghdad Abrar had found a job at Iraq's Ministry of Higher Education, where her uncle was a senior civil servant. In the six years since Saddam's ouster, this ministry had remained a haven for Sunnis, its dilapidated hallways a refuge for the scholars, academics, and civil servants whose families were like the al-Kubaisis—people with generations of academic credentials imprinted in their DNA. Before 2003, they had been the men and women behind the scenes who were keeping the government running. But al-Maliki's government paid little attention to it, as few of its personnel voted for his party.

Abrar went to work each day as a new hire in the department that oversaw Ph.D. thesis submissions for chemistry and biology, both subjects she was passionate about. Although she hadn't completed her own graduate work at the time, Abrar tried to fit in. She enjoyed the rigor of preparing graduates for their degrees and testing them. Still, she had less in common with her colleagues than she might have thought. Many of the other women in her department were already married, and several were Shiite. And when it came to the topic that was closest to her heart, her sister's death, she found she couldn't speak to her colleagues about it—the wound was still too raw. So she kept to herself.

Abrar poured more and more of herself into her internet avatar. Online, Bint al-Iraq, the Daughter of Iraq, was a loquacious, bold persona with plenty of confidence and friends. It was through her that Abrar had finally found her voice. At night, when the rest of the family was asleep, she would log in to one of her favorite chat rooms, Shumukh al-Islam. By 2010, the population of the forum was close to twelve thousand and discussions hewed to a very narrow line—jihad, the illegal American occupation, the crimes of the occupation, and Islamic jurisprudence. It was here that Abrar found people who understood what it felt like to lose a relative to the American crusader army. They felt her pain, and she started to understand theirs.

The only real jihad is the jihad of Al Qaeda, one post read. Killing the crusader is a divine pursuit, the righteous path for all real Muslims. Shiites, those infidels, should also be killed. Bint al-Iraq promptly agreed. The only good Shiite is a dead Shiite, she wrote.

Chapter 10
Hunting for Prey

Abu Ali al-Basri never had the constitution for gambling. Growing up, he watched men in coffee shops play cards or backgammon for a friendly wager. His father shunned such sport, saying only the weak believed that their fate could be improved by games of chance. Abu Ali saw the wisdom in these words when a rumor went around the neighborhood about one of his uncles who kept losing in cards. Those to whom the uncle was indebted were emboldened to criticize him in public, while his uncle became a magnet for ill fortune and lost his confidence.

Abu Ali woke up each morning with a steadfast belief that God would not part the heavens on behalf of man unless he worked to help himself. He had survived two decades in exile on the run from a dictator because he had used his wiles and embraced patience and caution, instead of emotion.

Which is why, in retrospect, it was ironic that the mission in 2010 that cemented Abu Ali's and the Falcons' reputation with the Americans as a trusted counterterrorism partner came about when the spymaster threw his normal caution out the window and promised to capture one of the top Al Qaeda operatives on the U.S. most-wanted list.

Through most of 2009, the Iraqi leaders of Al Qaeda had pulled off a string of extraordinarily sophisticated and deadly attacks in the Iraqi capital that made a mockery of the American military's boasts that their vaunted troop surge had quelled the insurgency. In reality, the tens of thousands of ordinary Iraqis trying to help the nation off its knees were terrified of dying on their daily commute. That year, Al Qaeda operatives had bombed the Ministries of Foreign Affairs and Finance, killing 90 people. Then they targeted the Ministry of Justice and Baghdad's provincial council, killing 155. The next target was the area near the Green Zone where Iraqi government employees parked and walked to work each day, and where another 127 people were killed. Finally, on January 25, 2010, suicide bombers rammed truck bombs into three hotels where foreign correspondents, diplomats, and aid workers lived and worked.

With the CIA's man in Baghdad, Mohammed al-Shahwani, gone, the Americans were desperate for a new partner who could stem the bloodshed and unravel the mystery of how the terror networks in Baghdad worked. It was vital for both the Iraqis and Americans to achieve a semblance of security in the Iraqi capital, where nearly one-fifth of the country lived. From Washington's point of view, both the White House and the Pentagon wanted to reverse the hardening sentiment that the Iraqi invasion had become America's greatest foreign policy disaster since the Vietnam War. Prime Minister Nouri al-Maliki, meanwhile, knew he wouldn't last in office if the bloodshed didn't end.

So on a late wintery day, when Baghdad's sky flattened to a heavy gray pall, a team of six Americans from the Special Forces command responsible for hunting high-level terrorism suspects drove from the coalition headquarters a few blocks away to Abu Ali's low-slung cinder block bungalow, on the northeastern corner of the prime ministry complex, for their first meeting with al-Maliki's favored spy.

As the head of the prime minister's security, Abu Ali had spent three and a half years working with protocol officers from the U.S. embassy as well as diplomatic and White House security teams organizing official meetings with the Iraqi leader. Western intelligence officers and advisers knew him from his silent presence at the Baghdad Operations Command, where national counterterrorism operations were planned and executed. Only recently, however, had they found out that this unassuming man was running his own intelligence operations, with the prime minister's support.

The Americans sat together at the far end of the massive wooden table that was too big for the room next to Abu Ali's office. The Iraqi spymaster didn't know what to expect from the meeting, whether or not the U.S. team would treat him as an Iranian stooge and a potential enemy, given the years of rumors Mohammed al-Shahwani had spread. He walked into the meeting eager for the chance to change that opinion.

Rather than animosity, the Americans brought goodwill. They asked if Abu Ali would be willing to help them bring Iraq's top terrorists to justice.

In 2006, the Americans had killed the first leader of Al Qaeda in Iraq, Abu Musab al-Zarqawi, the man who had orchestrated the jihadi insurgency against American soldiers and the Shiite-led government with the goal of creating a puritanical Sunni religious state. Four years later, the terror group's new leaders were veterans of this war. Most Americans would be unfamiliar with their names, but Iraqis knew the identities of these men dedicated to destroying their nation.

Two of these men were Iraqis: Abu Omar al-Baghdadi, an elderly theologian who had taken over the group, and Manaf al-Rawi, a Saddam-era police officer who after the 2003 invasion had become responsible for all attacks in the Iraqi capital. The third man, Abu Ayyub al-Masri, was an Egyptian who had trained and lived with Osama bin Laden in Afghanistan. As he was the organization's minister of war in Iraq and one of the oldest veteran Al Qaeda figures still alive, the American government had put a $1 million bounty on his head.

For all their military prowess, however, the Americans had struggled to find these three leaders. In fact, in 2007, the U.S. military had even announced that Abu Omar al-Baghdadi was not a real person. Given the Americans' failures, and the false reports that Prime Minister al-Maliki's Force 54 had captured and killed Abu Omar, all sides were desperate for help finding their prey.

Tell me who you see as the biggest threat and I'll get him within a month, Abu Ali told the Americans, eager for their approval.

He had little insight into who was on the Americans' most-wanted list, but he was confident that the twelve men he had assembled would be ready for any challenge. With the Americans' virtually unlimited resources and the Falcons' local knowledge, together they could actually make a difference.

Answering Abu Ali's question that morning, the American special forces colonel gave his new Iraqi colleague just one name: Manaf al-Rawi, the man responsible for the bombings in the capital and its government institutions.

Abu Ali sat back in his chair and took a sip of tea while trying to keep a straight face. He had never played poker, but in that moment he understood the euphoria of being dealt a winning hand. A year earlier, Abu Ali had managed to recruit a mole inside Al Qaeda, and his agent had secretly recorded a meeting held in Syria between Al Qaeda and other Iraqi Sunni insurgents. The intelligence Abu Ali gained from that meeting included the names of leading members of Al Qaeda in Iraq. One of the names that the agent had passed on was Ali al-Zawi. Another was Manaf al-Rawi, who had survived the battle for Fallujah and, later, three years in a U.S. military prison, all the while rising through the ranks of the terror group.

Abu Ali al-Basri had details about al-Rawi's activities. Now all he had to do was find him. If the Americans came to test him and his team, he wasn't about to lose.

One month, Abu Ali told the Americans. Just give us one month.

Most counterintelligence officials around the world will concede a fundamental fact about their job. Most missions are a combination of luck and hard work. Agents can provide reams of data. Analysts can spend days sifting through the information. Tactical teams can pinpoint exactly where the target should be during an operation. But until that person is caught, an element of uncertainty always exists. Anything can go wrong. The target could leave his safe house a few minutes before the arrival of an arrest team.

A couple of days after his meeting with the American colonel, Abu Ali had a stroke of luck. He learned from his Al Qaeda mole that al-Rawi was expected soon in Baghdad. Iraq was preparing for another election, this time for the national parliament. With al-Maliki's government again under pressure, Abu Ali knew he had a rare chance to get cooperation among Iraq's rival security agencies. The commander in chief would not want a huge bombing to undermine his chances at the polls. Abu Ali easily got approval for the special units controlled by the prime minister himself and the domestic intelligence agency—whose leader was an old friend from exile—to erect new checkpoints around Baghdad's outskirts and on the capital's major thoroughfares. The brigades manning those posts were given a poster with al-Rawi's face and name.

Despite the high alert, they failed to catch the Al Qaeda commander. Instead, terror cells detonated a hundred bombs and mortars throughout Baghdad on election day, killing at least thirty-eight voters.

Abu Ali refused to give up. The capital remained on high alert as politicians fought over the election results. Al-Maliki and his election coalition came in a close second to a rival bloc and the Iraqi leader was mounting a legal battle to keep his position. Amid the political turmoil, the Falcons made sure that al-Rawi's mug shot remained at the checkpoints. Four days later, on March 11, the vigilance paid off.

On Baghdad's northern edge, a federal policeman in the Hay al-Hattin neighborhood pulled over a dusty Chevy Caprice and demanded IDs from the passengers. Al-Rawi was traveling under an assumed name, but there was no mistaking his face. The terror suspect was taken into custody and handed over to Force 54.

That's when Abu Ali's work got complicated. Prime Minister al-Maliki ordered Abu Ali to keep the arrest secret from the Americans. Al-Maliki didn't want anyone except his own security forces to claim a counterterrorism victory, especially now with his political future on the line. Al-Maliki wanted al-Rawi to spill his secrets about Al Qaeda networks in the capital and he wanted results fast.

But for the next twenty days Force 54 couldn't get al-Rawi to crack. It's unclear how the unit treated their detainee and what methods they used to try to extract information from him. But al-Rawi didn't speak at all. Without any breakthroughs, al-Maliki's frustration grew and at the end of March Abu Ali took over the interrogation.

When he first saw the prisoner, Abu Ali felt sick to his stomach. A swarm of black flies darkened the meager light inside al-Rawi's cell. The bare concrete walls were filthy with rust-colored stains and dried feces. There was no toilet, no bucket, and no bed. Al-Rawi sat in a corner, his head between his knees, curled into a ball as tightly as he could. Abu Ali saw right away that the only tactic tried on the prisoner had been brute force. That, he thought, needed to change, now.

He walked into al-Rawi's fetid cell with his own strategy.

Boy, why is this cell so dirty? Abu Ali shouted at the guard. Why has the doctor not seen this prisoner? Who is responsible for this abuse?

The guard appeared bewildered. This type of concern for a prisoner was rare at Force 54's detention center.

That was just what Abu Ali thought would happen. If pain hadn't broken al-Rawi, then restoring his dignity might bring results.

Boy, I want this cell cleaned immediately. And I want this prisoner taken to the chief's room so he can shower, Abu Ali barked at the guard.

An hour later, al-Rawi was clean and wearing a freshly laundered dishdasha, a long, ankle-length shirt, and warm-up pants, clothes that the guards had told him were from Abu Ali's own wardrobe. Al-Rawi was given privacy to pray and then served tea, hot bread, sweet clotted cream, and honey, the favorite breakfast of Iraqi children.

Abu Ali entered the room again. He had al-Rawi just where he wanted him: grateful, warm, and with a full stomach.

The Falcons' chief then spent four hours asking Rawi about his family, how he grew up, who his father was, and how his mother had been faring since her son had abandoned her for Al Qaeda's jihad. He massaged the cultural muscle memory of every Iraqi man, the one exercised every day since birth—that sons are responsible for their mothers' welfare. Abu Ali assured him that if al-Rawi cooperated, Abu Ali would take care of his family.

The ploy didn't work. Al-Rawi sat across a small table not much larger than a teacher's desk and told Abu Ali that he had the wrong man. He said he was a bricklayer looking for work in Baghdad and knew nothing about Al Qaeda or explosives or terror attacks.

Abu Ali sat quietly across the desk, tapping his forefinger and trying to remain patient. He knew he had made some headway. The fact that al-Rawi was talking was already a breakthrough. But Abu Ali's deadline with the Americans was looming. It had been a month since his meeting with the colonel. Technically, he had won the bet they had set. Abu Ali had located al-Rawi. But he didn't have any useful intelligence to share with the Americans.

Abu Ali knew that although al-Rawi was behind bars, his network was still active. For all he knew, Al Qaeda had imminent plans to attack Baghdad. Abu Ali needed help, but al-Maliki, obsessed with his political legacy, was still adamant that the Americans not be told of the arrest.

Al-Maliki's stubbornness was notorious. This time, it had deadly consequences. On April 4, with the election results still unresolved, Al Qaeda suicide bombers rammed cars loaded with explosives into the Iranian, Egyptian, and German embassies in Baghdad. Another bomb targeting the French embassy failed to go off and Iraqi security forces succeeded in arresting the bomber. At least forty-one people, all of them Iraqis, died in the blasts. The tragedy was the last thing that al-Maliki needed.

While emergency services were still counting the casualties and Iraq's international allies were making angry phone calls to the prime minister's office, Abu Ali pressed his boss to compromise. We need to bring the Americans in now. There is no other choice, he told al-Maliki. The prime minister finally agreed.

Abu Ali walked across the courtyard from the prime minister's offices to his own office and contacted his American colonel.

Abu Ali told him that he had al-Rawi in custody, and the Americans had some surprising news for him in return—the Americans had detained al-Rawi's twin brother. Abu Ali knew that this was exactly the type of leverage he needed to overcome al-Rawi's resistance.

The spymaster knew that if you were going to persuade a man to betray a cause, to turn his back on the life he had been living for seven years, and disclose the names of the men he had fought with, you needed to give him an honorable rationale for that treachery. For an Iraqi, that reason would be his family.

Abu Ali told the Americans that if they agreed to trade the twin brother's freedom for information about the Baghdad terror networks, then Abu Ali could get al-Rawi to open up. After spending several days with the detainee, the Falcons' chief had concluded that al-Rawi was not a radical religious ideologue like the foreign fighters Al Qaeda sent to Iraq. The man was a murderer and a traitor, but for Abu Ali's purposes, he wasn't completely irredeemable.

The Americans decided to gamble that Abu Ali knew best.

They handed al-Rawi's twin over to the Iraqis. Then, the Falcons' chief started to twist the emotional screws he knew no Iraqi man could resist. He put the brothers in the same room and called their elderly mother. He explained to her that she could lose both sons because their alleged crimes deserved the death penalty, or Manaf could save his brother's life. All he had to do was tell Abu Ali what he knew about the Baghdad terror networks. Listening to his mother's tears, al-Rawi finally broke. The stream of names and details that he provided about safe houses, bombers, and supply chains took days to take down. When he finished, Abu Ali had Al Qaeda's complete organizational structure in Baghdad.

As Iraqi federal police began arresting suspects around the capital, Abu Ali realized that the rich vein of intelligence that he had tapped went even deeper.

Once he cut the deal to save his brother, al-Rawi shed his reticence. He gave Abu Ali the most precious nugget of information. When pressed to explain how he chose his targets in the capital, al-Rawi told the spymaster that decisions were made after consulting the two Al Qaeda men in Iraq higher on the most-wanted list than he was: Abu Ayyub al-Masri and Abu Omar al-Baghdadi. A single courier sent letters between the three men.

The two Iraqi leaders lived off the grid, never used phones or email for communication, and relied solely on couriers delivering handwritten letters. The Americans, who had been using their high-powered electronic surveillance infrastructure to vacuum up data from Iraq's telecommunications networks, knew the two leaders had no digital footprint. For four years they had been trying in vain to find them. Now, Abu Ali had found a solid lead.

The information led the Falcons to a two-story home near Baghdad's historic Bilal al-Habashi Mosque, a discrete quadrant in the sprawling metropolis favored by middle-class Sunnis and not Al Qaeda supporters. But there, on a small lane bisected by tall concrete barriers protecting an adjoining school from possible bombings, lived the courier who for months had been the link between Al Qaeda's two leaders in Iraq and the outside world, a man named Jaafar.

When Abu Ali's men brought him this news, his stomach churned with excitement. Al-Rawi's capture had remained secret, but he couldn't be sure that the Al Qaeda leaders hadn't noticed his absence. If they had, then they might have ended the courier's deliveries, out of an abundance of caution. If they hadn't, then the Falcons needed to move quickly and discover whether the path that could lead them to Iraq's two most-wanted men remained warm.

Abu Ali ordered a stakeout that same night, putting his oldest officer in charge. Major Bassam, a tall, lanky chain smoker with a thin mustache that curved upward like a smile, hailed from southern Iraq and had also lost his father to Saddam's mukhabarat. Abu Ali chose him to be his deputy because, unlike many others who had filled the copious jobs in Iraq's security services, Bassam understood that the best intelligence wasn't obtained with a series of kidney punches to a detainee. Instead, brains, patience, and manipulation were the ingredients for success.

Shortly after midnight, Bassam, dressed in a long-sleeve black shirt and armored vest, drove with a team of four men in dented, unmarked pickups into the neighborhood. The Falcons hadn't had time to set up proper surveillance of the street, and they didn't know the courier's routines. They hoped that he would either be home asleep, or that someone from his family would be able to tell them where he was.

When Bassam's men crept up to the home, the building was dark. They picked the lock and looked inside. Dinner dishes were soaking in the sink and the bed pillows were warm, but no one was there. Bassam wasn't sure what to do. He had a hunch that his quarry was still within reach. He would have to be patient. The team would settle in for the night to watch.

As Bassam stepped out of the house he saw car headlights pulling up the street. He hid behind the front door, cautioning his men to stay quiet. Outside, a car door opened and then shut, and two voices broke the stillness. By the familiar way they talked, the man and woman were clearly a married couple. The wife was complaining about intestinal pain and her doctor. The husband was asking her to be patient while he parked the car. Bassam heard the car door open again and decided to act. If the man speaking was the courier, he couldn't give him a chance to drive away.

Go! Go! Go! He yelled to his men and ran through the front door.

The woman screamed, frightened by the rush of bodies flooding past her. One of Bassam's men leapt and grabbed the driver in a bear hug. The suspect was off balance, caught halfway out of the car, but he fought with the desperation of the doomed. He shook off his tackler and started running away. Bassam's team opened fire, shooting him four times before he went down. Stop shooting! We need him alive, Bassam shouted over and over.

The veteran officer got a tourniquet on the man and rushed him to the hospital. Bassam stayed in his room all night, smoking and pacing by his bedside.

By the time the sun rose, the Falcons were confident they had gotten their man. They knew Jaafar's full name and the number of relatives who were already in Iraqi custody, including the patriarch of his family. When the courier awoke, Bassam stood by his bed with that elder man, Jaafar's uncle, by his side.

Salaam wa Alaikum, the Falcons interrogator told him. You should thank me. I saved your life.

Jaafar was handcuffed to his bed, weak from a loss of blood. But his wounds did nothing to moderate his anger. Son of a bitch. You tried to kill me before saving me, he said, spitting at Bassam.

The major laid out the case against Jaafar without any pretense or tricks, letting him know just how much the Falcons already knew about the network and how helpless his situation was.

As Bassam spoke, Jaafar lay with his eyes closed, giving no sign that he heard the words. His uncle was pale and visibly shaken. Bassam then offered Jaafar a lifeline.

He told him that if he supplied enough information to lead them to Al Qaeda's top men, then Bassam would free Jaafar's son from prison, where he was scheduled to be executed. He would make sure that Jaafar would receive a life sentence for his crimes, instead of the death penalty. And that the rest of the family would have immunity from future prosecutions.

At this, Jaafar's uncle interjected. The amnesty meant that dozens of relatives would have the opportunity to live normally and not be targets of retribution, like so many Iraqi Sunnis, for being related to convicted terrorists. He ordered Jaafar to accept the deal—and start talking.

By lunchtime, Bassam learned more about how Iraq's most wanted men communicated with each other than the Americans had in four years. He knew the courier's schedule, how the terrorists hid messages under the soil of terra-cotta flowerpots, and the final link in the communication chain, the house in Samarra where a second courier lived.

He also learned that Jaafar's next mission was due to start in seventy-two hours. We only have three days to plan our operation if we want to find the top two terrorists in Iraq, Bassam told Abu Ali, back at his chief's office.

Bassam then said something they both knew was politically inconvenient but nonetheless true. The small Falcons team might have gained a treasure trove of intelligence, but Abu Ali had only a dozen men under his command, barely enough for a twenty-four-hour stakeout of a house in Baghdad, a city they all knew like the back of their hands, let alone Samarra.

If they wanted to succeed in nabbing the country's most wanted men, they would need to call their new American friends.

For the second time in a month, and for the second time in his life, Abu Ali contacted the American military. His boss, Prime Minister al-Maliki, wouldn't like what he planned to do, so he didn't inform him beforehand. Abu Ali didn't care about politics. He cared about results. It was another unusual gamble for the normally pragmatic spy chief, but the tight time frame left Abu Ali with little choice.

Abu Ali sat on a metal office chair in a classified operations room hidden in a corner of Camp Balad, a

U.S. military base outside of the Iraqi capital. He was concentrating on keeping his facial expressions calm, fighting the urge to fiddle with his eyeglasses. Three days earlier, he had felt confident with their chances, but now his mood had flipped. He had what his American colleagues called pregame jitters.

When he gave his first briefing to the Americans, their skepticism was evident. He didn't fully understand the mad scramble his information had set off between Baghdad and Washington. He didn't appreciate how rare it was for an Iraqi official to be allowed into the hub of American military operations. He didn't know the Americans had committed hundreds of millions of dollars to the operation, from jet fighters to surveillance drones to highly trained men. He didn't know that the White House itself had been briefed about the mission to find and possibly kill the two terrorists responsible for killing dozens of U.S. servicemen, international aid workers, and Iraqi citizens.

What he was acutely aware of, however, was that if the operation succeeded, then his boss, Prime Minister al-Maliki, would take full credit. But if anything went wrong, then Abu Ali, sitting hundreds of miles away from the action, cut off from his men and surrounded by screens showing live feeds of the target, would alone be blamed. Yet what consumed Abu Ali in the moments before the operation began was not the fear of failure, it was the burden of moral responsibility. His Falcons and the Iraqi special forces, along with their American allies, were risking their lives on an operation that he had put into motion.

Operation Leaping Lion started as planned, before sunrise on Sunday morning, April 18, when Jaafar the courier set off from Baghdad with a cargo of potted flowers.

He drove north to Samarra, the same town where in 2006 Al Qaeda had blown up Iraq's Shiite shrine. The city, about eighty miles north of the capital and close to Saddam Hussein's birthplace, had remained a key hub of Sunni insurgent recruiting and activity, in part because it was the home base of another senior member of the group, Abu Bakr al-Baghdadi, the man who would later become the head of the Islamic State.

The Americans kept aerial surveillance on the car, following it to a building on the edge of the city where the driver stopped and went inside. This was where Jaafar would meet the second courier and pass the letters onward to him.

The American satellite feed showed a second vehicle arrive, and a second driver got out of this small pickup. After a quick meeting inside the building, the second man transferred the flowerpots from Jaafar's vehicle to his. In the command room, Abu Ali watched the footage from a drone following the moving cargo. The courier first traveled westward before doubling back, like a man afraid of being followed. He stopped at a hardware store and bought sacks of concrete mix. He then stopped at a wholesaler and bought flour, throwing those sacks on top of the flowerpots.

When he got moving again, the driver left Samarra then doubled back once again. He drove to a used car lot and tried to swap his truck for another vehicle. Apparently, though, he couldn't strike a deal and after a while, he got back on the road in the same pickup truck.

Meanwhile, a team of American and Iraqi special forces had mustered in Samarra, ready to launch a ground raid once the courier reached his destination. By afternoon, the pickup had reached Thar-Thar, a desert area west of Samarra that Al Qaeda had used for training camps. The team in Camp Balad watched the truck pull down a rutted lane to a farmhouse with a thatched roof and no one in sight. The isolated location was a perfect hiding place. No one could approach without the people inside knowing. At the same time, though, there was nowhere to escape.

The American commanders gave the green light for ground forces to approach, confident that this was the hiding place they had spent years searching for.

Iraqi special forces from the 54th Brigade and U.S. commandos surrounded the farmhouse. Sixteen people, including the wife of Abu Omar al-Baghdadi and several of his children, surrendered. A thorough search of the house didn't reveal any sign of Iraq's most wanted men.

Back at Camp Balad, Abu Ali started sweating when he heard this news. His intelligence now resembled an overripe grape, plump and attractive until you taste its sickly rotten juice. He walked to the corner of the command room and in a rare moment of vulnerability called Bassam, who was with the ground unit.

Brother, we have to be missing something, Abu Ali told him. We can't allow everything to turn to shit.

Bassam told him to talk to Jaafar. The time for playing good cop was over. Clearly their detainee had not told the Falcons everything he knew.

The spymaster told Jaafar that his deal was off the table. We will kill your son and kill you unless you tell me what has gone wrong, he screamed.

Jafaar finally divulged the last crucial clue. The farmhouse in Thar-Thar had a hiding place, under the kitchen floor. When the Iraqis relayed this to the Americans, the commandos tore open the tiled floor to find the two top Al Qaeda leaders in Iraq. Abu Ayyub al-Masri and Abu Omar al-Baghdadi died in the brief gun battle. The Americans retrieved a large cache of files, including the pair's communications with top Al Qaeda leaders in Pakistan, and blew up the house.

When news of the operation reached the White House and the Iraqi prime minister, the Americans and Iraqis pushed ahead with more operations based on this fresh intelligence. Over the next three days, Operation Leaping Lion captured or killed another dozen Al Qaeda senior figures, including many in the northern Iraqi city of Mosul. For the first time since 2003, the head of American forces in Iraq, General Ray Odierno, and the White House publicly congratulated the Iraqi security forces for the counterterrorism victory. Prime Minister al-Maliki basked in the adulation, while Abu Ali contented himself with quiet handshakes from his American counterparts.

The nature of our work is to keep to the shadows, they told him. No one may understand your role in all this, but rest assured that we do.

They left him filled with optimism.

You know how to reach us, they said. Whenever you have another tip, let us know.

Chapter 11
Living Your Best Life

In the fall of 2010, Harith pecked at the keyboard on his desk tucked away in an office down a long, dim corridor at the Ministry of Electricity. His outmoded IBM computer screen was partitioned into four separate video feeds: static images from cameras placed around the power plant in Dora, on the southern edge of Baghdad.

How has my life come to this? he asked himself. A boy who once composed poetry, the young man who felt the stirring of romantic love, had been reduced to a life as an office drone, sitting alone monitoring an unchanging camera feed.

For many in Iraq, the year so far had delivered terrific progress. Prime Minister Nouri al-Maliki had prevailed in an acrimonious and controversial election, and renewed Abu Ali al-Basri's mandate to fight terror thanks in large part to the success of Operation Leaping Lion. Munaf al-Sudani had joined the Falcons and had spent months training on surveillance equipment used to eavesdrop on terror suspects. But Harith's boss, the minister for state security, switched jobs after the election, and his office staff, including the oldest al-Sudani brother, were out of work. Harith's recommendation from the minister was solid, but he had few job options. That's how he ended up, at age twenty-nine, in a desk job as a junior officer at what, on paper, was one of Iraq's new and vital security agencies—the one tasked with protecting the country's oil pipelines and electricity stations. For years, Iraq's electrical power stations had been a favorite target for insurgent attacks. Al Qaeda knew that if the Shiite-led government of the world's third-largest oil supplier couldn't keep the lights on for its citizens, then the group would have a wedge issue to use, especially among Iraq's Sunni communities, to foment unrest against their authorities.

Although the ministry had tremendous responsibilities, the attitude of most of Harith's colleagues was lackadaisical. They approached the job like typical Iraqi civil servants. It was not even four P.M. but the office was empty—everyone had clocked out and gone home. They arrived at the ministry sometime in the mid-morning, drank tea in small groups around a samovar tended by the janitor, and left as soon as the sun was high. It didn't matter that they were all uniformed members of Iraq's security forces. They weren't spies, like his younger brother Munaf, someone with rank, someone destined for exciting missions. Unlike his colleagues, Harith tried to think of his job with a sense of pride. But it was hard to escape the truth that what they were doing as guardians of the country's critical infrastructure was simply drinking tea and watching video screens.

In 2009, youth unemployment in Iraq was a soaring 18 percent. If a man had no diploma and no political contacts, he was doomed to struggle, like Harith's cousins who hauled cargo at the market, or join the lethargic army of jobless men bunkered down at the city's cafes who discussed politics or soccer, or nothing at all, stretching their small pot of tea into midday. In such an environment, any government job, however tedious, was a blessing, as it meant a salary for life.

But Harith was no plodder. He had spent too much of his life aimless, and now he had an itch to be better, to do something more. Wasn't that what he had been told his whole childhood? While other colleagues were wasting time, Harith sought opportunities to elevate himself out of the job he didn't like.

That's why he remained at work later than everyone else at the tall, dingy ministry building in Baghdad's western neighborhood of Mansour. Alone in his office, Harith spent hours searching the internet, something he couldn't do at home. The al-Sudanis didn't own a computer, but even if they had, with the electricity blackouts in Sadr City, and the ruckus of his own small children and the dozen or so other relatives there, he never would have found a moment of peace. While he was online, he had a window to the rest of the world. He could see how other militaries worked; the tools that police in Europe used in their jobs. He could see beautiful places unlike anything in his wildest imagination. The blue waters of the sea and snow-covered mountains.

On the rare nights when Harith was in Sadr City, he preferred the company of his brothers to that of his wife, Raghad. The men would spend hours together playing video games, and she would fall asleep alone. The mornings, when she got the children ready for school, were among the few times they were all together. She tried not to show her bitterness to Harith, but usually, she failed. Her husband was the toughest trial that God had ever sent her.

Your daughter fell asleep crying again last night. They wanted their father, but you weren't home. Don't you have any shame? Causing them to suffer so?

When Raghad started nagging, Harith simply walked out of their apartment and went downstairs, where his mother prepared breakfast for the family. Like many Iraqi women, Raghad was brought up with the idea that to be respectable, she must never lose her composure. Raghad might know the family's own unresolved issues with Harith, but her role in the family was to demonstrate unwavering support for him. To say a word against her husband in front of the other al-Sudanis would be unforgivable.

One bright spot in Raghad's life, and in all the al-Sudanis' lives, occurred earlier in the year, when Munaf announced his recruitment by an elite intelligence unit. The news lifted the gloom over the house, as each of her in-laws began recalculating how Munaf's promotion would affect their family's social standing and the job possibilities for his other brothers. Raghad had her own reasons for hope. She and Munaf's wife had become close friends, and Raghad knew her own position in the family would rise along with Nismah's.

Harith had been the model of tact after Munaf's announcement. The spat about the crushed minivan was well behind the two brothers. Despite their father's counsel, Munaf had slowly repaid his brother for the repairs. Besides, it was a relief to Harith that their father had someone else to complain to whenever there was a problem.

Harith invited Munaf out to their old cafe, the dingy hole in the wall near the police academy in central Baghdad, to smoke shisha and see if he could glean something to apply to his own career.

Munaf later remembered the evening as a turning point. Harith let Munaf walk into the teahouse in front of him and choose which seat he wanted at the table. When the boy brought their water pipes and placed the red-hot charcoal in the tray to heat the tobacco, Munaf could see subtle changes in his brother. The anger that had consumed him for so long had disappeared. In fact, Harith had a heartfelt smile on his face when he congratulated Munaf on his good fortune.

While they were growing up, Harith never swaggered at home. Everyone could see that he was clever, but Munaf wasn't sure Harith believed in himself. His father's beatings and berating had badly wounded his oldest brother's sense of confidence. Munaf knew that Harith had not had it easy, but at the same time his big brother had never appreciated what it had been like for the rest of the al-Sudani brothers to live in his shadow—the way they all had to clear a room when Harith needed to study, the obvious implication that somehow, intrinsically, the other ten al-Sudani children were of lesser value.

As he drew the tobacco smoke into his lungs, Munaf could see that for all those advantages, Harith still needed help in seeing what the rest of the family had all been raised to believe: that he could succeed at whatever he put his mind to.

What is it that you really want to do? Munaf asked his brother after Harith told him about his frustration at his new security job. Harith said that he didn't know. He asked what kind of work his brother would do as a Falcon. When Munaf described how their mission was to track and find the country's most-wanted terrorists and keep the nation safe, his older brother's eyes lit up.

Munaf told Harith that Abu Ali al-Basri had tough standards. His standing at the top of his class at the academy had put him on the spymaster's radar. But good test scores were not enough. He wanted men who could think on their feet, who could work long hours and take calculated risks, like Munaf's trips to Sadr City. You need to find a way to get some experience, find someone you trust who can help you, he told his older brother.

Harith's fist clenched around his water pipe. The only person he knew in the intelligence field was his brother. After a few moments of reflection, he remembered someone else, a senior officer in his own security force. When Harith first started his job, the older man, Colonel Ali Hussein, had lectured the new men about operational security and the importance of keeping secret information they learned on the job. Colonel Hussein oversaw intelligence-gathering for the force, and although Harith didn't know anything about his background, he was impressed with the senior officer's bearing.

By the time the brothers returned home that night, Harith had already started thinking about how to ingratiate himself with Colonel Hussein. If moving up in life meant taking risks, then this would be his first step.

Colonel Ali Hussein stood in the darkened well of the auditorium full of young guardsmen. They were packed into sleek modern desks rising in a crescent shape above him and his video monitor and listening to his lecture about the ministry's new technology.

As the head of intelligence for Iraq's oil infrastructure and power stations, he had the monumental task to keep these critical assets safe from sabotage or bombing. The colonel noticed many heads nodding down toward their owners' chests, something that happened quite frequently during his tutorials. He also noticed that one man in the center of the room was writing down the points that he had emphasized.

When he finished his lesson, the room was silent. Hussein wasn't sure anyone had stayed awake. Except for that one eager man who was now moving swiftly down the aisle toward him.

Ya siidi. Your excellency, Harith called to him, snapping a crisp salute. I wanted to thank you for your lesson today. Can you tell me more about the operational precautions that you are taking? I'm very curious about intelligence matters.

Colonel Hussein didn't know whether to slap the young man for his impudence, or to smile. Who did he think he was, approaching a senior officer in such a manner? Anyone who grew up in Saddam's Iraq must know that no one in their right mind would approach an intelligence officer with such a request.

Harith was persistent, pestering the colonel until he elicited a promise for a meeting the next day. Such eagerness riled the colonel. Anyone that earnest must be simpleminded, or hiding another agenda, he thought. Harith hadn't been a member of the guard force for long. Maybe he was a spy sent by a rival agency to assess how the colonel ran his department? Colonel Hussein decided to monitor Harith himself and get to the bottom of it.

Despite his wariness, the intelligence officer found himself charmed during his meeting with Harith the following day. Something about his demeanor, the officer thought. Harith was two parts fawning but also two parts sincere. The colonel's own team reported that they didn't find anything untoward about the background of the man from Sadr City. He was clever and a hard worker. He'd finish a task in a third of the time it took another junior officer. The family had never been suspected of political crimes. He was as clean as an Iraqi man could be.

Although it broke all professional protocol, the colonel decided to give Harith, who had no formal intelligence training, small support roles on operations that the colonel ran against saboteurs of Iraq's energy pipelines.

Harith absorbed information like a sponge. No one could have asked for a better apprentice, the colonel thought. He recommended Harith for government-funded classes in computer coding, eager to help the young man get ahead.

Within a year, Harith was selected for training at Camp Dublin, the U.S.-run program at an Iraqi military base southwest of Baghdad. He passed the course with distinction and a month later returned to the ministry a changed man. Harith had adopted the style and demeanor of his American trainers. He had started wearing a Bluetooth headset for his phone and the reflective Ray Ban sunglasses common among GIs. His colleagues started calling him Robot because, they joked, he had fallen in love with technology, just like the Americans.

Harith always seemed to be one of the first people in the office each morning, and the last to leave in the evening. Like his days at the Jamila market, where he had acquired the reputation of being able to sell carpets to a rug merchant, Harith had worked to get what he wanted. A way out.

Chapter 12
Alone in the Wilderness

In June 2012, it had been six months since the last American troops left Iraq, and Abu Ali al-Basri was still trying to cope with the feeling of abandonment. In Washington, political leaders were busy spinning the withdrawal as a mark of success, a claim echoed by his own leaders in Baghdad. Both had declared victory over Al Qaeda, announcing that the relentless military campaign to eradicate the group's supporters in the Iraqi Sunni regions in the west and north had broken the hold the terror group had on the country. The intelligence breakthroughs that had eradicated the sleeper cells in Baghdad meant a job well done.

Such assertions had a ring of truth, and they certainly made for great speeches for politicians, but Abu

Ali and other security professionals treated them as haki fadi—empty talk.

Washington may have had the luxury of shifting their attention away from Iraq to other national security concerns, but Abu Ali did not. His nation was still in Al Qaeda's crosshairs.

Just before the last Americans departed in December 2011, the Iraqis organized a joint counterterrorism raid, relying on information gleaned from one of Abu Ali's most trusted double agents, a jihadist in the northern city of Mosul. The two-day operation netted a dozen Al Qaeda suspects and, more crucially, a piece of intelligence that had increased Abu Ali's concern. The Falcons found a list of twelve hundred names—the vast majority of them Iraqi veterans of jihad and ideologically driven. The fact that they had evaded the U.S. dragnets and were still at large meant that they were a real and present danger for Abu Ali.

The grim reality after the formal exit of American combat forces was that Abu Ali was operating with fewer tools to keep his nation safe. The poisonous atmosphere of Baghdad's bureaucracies had increased in recent years, as Prime Minister al-Maliki fostered a climate of mistrust that made security commanders risk-averse and unwilling to share information.

The lack of political urgency in Washington meant the Americans' vaunted electronic surveillance networks around Iraq went unmonitored, and Abu Ali and his Iraqi colleagues had fewer partners to call on. The number of U.S. Special Operations and CIA teams focused on Iraq dropped sharply.

The Iraqi spymaster, like most of his colleagues, had not had the luxury of pursuing higher education and lacked the technical skills that their American counterparts took for granted. Abu Ali was a man who preferred paper files, took handwritten notes or committed hours of meetings to memory, instead of recording them or typing them in real time on Microsoft tablets, like many of the Americans he had worked with in the past five years.

His approach to intelligence-gathering was similarly traditional. Abu Ali relied on human sources rather than digital know-how to build his counterterrorism prowess. Al Qaeda's top man in 2012, Abu Bakr al-Baghdadi, knew of the American penchant for high-tech, and the reason he and his top echelon of commanders had stayed out of the U.S. military crosshairs for so long was that they shunned electronics. Besides, they were of a generation who, like Abu Ali, were too old to learn new digital tricks. They would only trust what they heard face-to-face from valued interlocutors, not what they were told over a digital phone screen.

Still, the Iraqi spymaster knew he couldn't afford to be too far behind the times, which is why he had sent a quarter of his men to train on the telephone and other electronic surveillance equipment that the Iraqi security forces had bought after 2003. He wanted a unit of the Falcons skilled at tracking militants online, the virtual recruiting zone where Al Qaeda was gaining converts.

Among the men assigned to the Falcons' electronic surveillance team, Munaf stood out. He had become adept at monitoring jihadi websites to build data banks of members and the links between disparate networks. So in the summer of 2012, when Abu Ali received more funding to expand the Falcons' ranks, it was natural that he turned to his bright lieutenant for suggestions about other recruits like himself.

The conversation was brief, a meeting that took place after a particularly long shift for Munaf. Abu Ali sat at his wide, engraved wooden desk that as usual was completely hidden by piles of manila and green official folders that seemed one deep breath away from toppling over. My son, the spymaster asked him. Who could you recommend to help bolster the ranks of our cyber team?

Munaf's reply was succinct. Chief, I know just the man for you.

A week later, Munaf ushered his brother Harith into Abu Ali's office. The brothers knew the moment was portentous. A job with the Falcons was the possible golden ticket that the eldest al-Sudani had been hoping for.

Munaf had called Harith right after his conversation with Abu Ali and told him to meet at their usual cafe by the police academy. Even in the smoky dim light of their corner table, Munaf could see his brother's eyes light up. He kissed his younger brother on both cheeks and asked him what he needed to do to prepare for an interview.

Abu Ali had already requested recommendations from Harith's former employers, so Munaf knew his brother was a serious candidate. He advised Harith to prepare to talk about his most successful security operations and work experience. Harith vowed to fast until the day of his interview so he could lose weight and look fit.

The night before the meeting, Harith asked Raghad to iron his dress uniform and told his son Muamal to shine his shoes. He then went to his father and asked him to pray for him.

The next morning, as they drove to the interview, Munaf had one last piece of advice. Don't be too cocky, he warned. You won't get the job simply because we are related. The chief isn't like that.

When the two al-Sudanis walked into Abu Ali's office, the spymaster sized up Harith. He already knew his reputation as a disciplined and diligent worker. As he introduced himself, Abu Ali could see from Harith's bearing that he had a calm confidence. He didn't fidget in the silence that the spymaster let hang in the room for several minutes, a tactic Abu Ali often employed in the interrogation room to size up a detainee. Abu Ali knew from the appraisals he had received from Harith's former bosses that the older al-Sudani worked well in a team yet also knew how to seize initiative. That was the kind of modern thinking that Abu Ali wanted. He needed new ideas to defeat Iraq's terrorists. When he asked Harith how the Falcons could be more proactive about tracking terror suspects online, the young man didn't hold back.

Twenty minutes into the meeting, Munaf relaxed. He thought his brother was making a good impression, but he knew Abu Ali never acted in haste, and they'd have to wait for his decision. To his surprise, as the interview came to a close, Abu Ali stood up from behind his messy desk and shook both brothers' hands. Harith, he declared, you're hired.

For the next three months, Harith undertook the intensive training that his brother had completed years earlier. He learned technical skills necessary for electronic surveillance and techniques to create aliases and cover stories needed to infiltrate the encrypted online chat rooms where Sunni jihadis congregated and discussed attacks. When the training was complete, Harith joined Munaf's team on the Falcons' cyber unit, joining his brother on one of the newest battlefields in the global war on terror.

Until Iraq became a focus of Al Qaeda's militancy after 2003, the puritanical form of Sunni Islam practiced in Saudi Arabia and honed by the terror group into an ideological sword was not indigenous to Iraq. Indeed, following the U.S. invasion in 2003, Iraq's Muslim majority, both Sunnis and Shiites alike, recoiled at the perverted expression of faith that the jihadis espoused in justifying the killing of innocent women and children. Iraqis' widespread revulsion for Al Qaeda coupled with a despair and anger at the government boiled over into popular protests against Prime Minister al-Maliki and his government. In December 2012 both Shiite and Sunni residents swarmed the streets of central Baghdad, calling for the leader to step down, mirroring protests across Iraq's Sunni-dominated north and west.

Although the demonstrations were mostly peaceful, the prime minister refused to heed demands for better infrastructure, jobs, and political rights. Al-Maliki denounced the protesters as terrorists and unleashed his security forces against them. The crackdowns took on a sectarian tinge. Iraqi forces killed dozens of Iraqis and detained thousands of Sunnis on terrorism charges.

Iraqi State TV depicted the demonstrators as bearded savages, but on the ground, the protests were populated with ordinary middle-class Iraqis, moderate religious sheikhs and politicians, poets, and even emotional, distressed grandmothers. Included among them were Abrar al-Kubaisi and her father. Professor al-Kubaisi, who on the weekends would have his son, or the al-Kubaisis' neighbor, drive them to the sprawling protest camp in Ramadi, which encompassed several city streets and had an almost carnival-like atmosphere. Coffee carts provided refreshments while vendors sold roasted nuts and fruit. An outdoor stage hosted speakers for the crowds, while dozens of tents arrayed around the perimeter became venues for sheikhs and notable politicians to hold court.

Before the Christmas holiday, the al-Kubaisis drove to Ramadi to attend one such gathering to discuss al-Maliki's recent intimidating speeches. Abrar soaked up the energy of the crowds and spent time volunteering at the first-aid center, while her father sat with his tribal elders.

When they returned to Baghdad, Abrar took home with her the frisson of excitement and satisfaction in finding kindred spirits. The protest camp had a vitality that her online chat rooms lacked. Online, her comrades were self-centered and competitive, like they were auditioning for a role in an ideological play. In Ramadi, Abrar found Iraqis who complained bitterly about what they saw as discriminatory behavior by al-Maliki's government, but they were polite and grateful, not arch. She could work for hours in the medical station in the blazing heat, and there was never a single soul who neglected to thank her for her help.

On Christmas Eve, al-Maliki made a nationally televised speech wishing the Iraqi people happy holidays. The address had been part of Iraqi political tradition since Saddam's ouster, a way for the government to honor Iraqi Christians, who comprised one of the largest congregations in the Middle East. At first, as he started speaking about national unity, it seemed that the prime minister would use the occasion to apologize for his government's heavy-handed response to the protest movement. But then, his tone changed. The prime minister's voice rose and he began castigating the Sunni protesters, branding them terrorists.

The Christmas address fueled the demonstrations, and more people took to the streets. Three days later, the prime minister ordered a raid against the largest of the protest sites, the Ramadi camp where the al-Kubaisis had spent time. The government assault killed at least thirty people, including old women and young boys.

Looking on in horror, Abu Ali didn't understand what had happened to the man he once knew as a pragmatic leader. Power, it seemed, had gone to his head, and, like Saddam before him, he had started seeing enemies everywhere. It didn't matter that Abu Ali had showed him the Al Qaeda membership list with twelve hundred names on it. The prime minister saw all Sunnis as a threat. At the end of 2012 Iraqi jails were overflowing with Sunni men despite any evidence connecting them to terrorism.

It was clear to the spymaster that al-Maliki's actions would sow more instability. Although the protesters weren't jihadis, Al Qaeda sleeper cells were ready to exploit the escalating unrest for their own ends. In the days that followed, Abu Bakr al-Baghdadi's emissaries showed up at the funeral prayers for those killed and commiserated with the disgruntled Sunnis. When the agitators whispered about plans afoot to topple the prime minister, few of Iraq's Sunni leadership thought of lifting a hand to stop them.

Chapter 13
Awakening the Beast

In early May 2014, Iraqi security forces arrested seven members of the militant group Islamic State in Mosul, Iraq's second-largest city, known for its Sunni businessmen and wealth.

It had been more than two years since the last American soldiers had left Iraq. The sectarian bloodletting between the country's Sunnis and Shiites had been mostly stanched, and the Western world had forgotten about Iraq, its troubles, its tragedies, and the terrorism that had nearly destroyed it. Indeed, American military commanders back in Washington, boasting that they had decimated the ranks of Al Qaeda in Iraq, were receiving promotions and lucrative post-retirement jobs. In Baghdad, Prime Minister al-Maliki was consumed with staying in power after what to him was another unsatisfying parliamentary election that revealed the deep divides of the country. His party had lost seats and his own hold on power was under threat. In his scramble to stay politically afloat, he paid scant attention to what Abu Ali al-Basri believed was one of the most important catalysts for terrorism: widespread arrests of Sunnis. In that fashion, al-Maliki mimicked his colleagues across the Middle East, where rulers believed that angry citizens should be pulled from the streets like weeds, instead of instilling democratic reforms.

Abu Ali, however, viewed such practices with disdain, believing that punitive security measures backfired, breeding more anger and terrorism, not less.

The truth was that although the rest of the world had shifted its attention away from Iraq, men like Abu Ali couldn't afford to. The cancer of extremism was real, but most of the Iraqi security forces were focused on the imagined enemy, all Iraqi Sunnis, instead of the Al Qaeda leadership and members verified in the group's own documents seized by the Falcons. Abu Ali knew that Iraqi lives depended on tracking this real threat, not the imagined one—even if it was just him and his Falcons, which now numbered four dozen men.

The political climate in Baghdad was treacherous, with Prime Minister al-Maliki supporting the wholesale roundup of prominent Sunnis across the country, wrongheadedly believing as he neared his tenth year in power that the sectarian affiliation of his critics made them a fifth column. Abu Ali, however, tried to keep his head down and focused on his specific targets, the veteran Iraqi jihadis who had remained alive and revived the organization after the Iraqis and Americans killed the previous heads of Al Qaeda in Iraq in 2009. The man who had taken the reins of power, Abu Bakr al-Baghdadi, had spent the years after the American withdrawal from Iraq restocking Al Qaeda's ranks with recruits from the Sunni areas of Iraq and Syria, the nation to the north that was convulsed by its own civil war. He rebranded the group as the Islamic State of Iraq and Syria, or ISIS, to reflect its new cross-border reach, and filled its coffers with cash from extortion, robbery, and smuggling. The Syrian war provided the new recruits a way to gain frontline battlefield experience, skills that the terror leader knew would be necessary before his planned strike in his homeland.

The men arrested in early May revealed a deep knowledge of the ISIS financial networks in Mosul, and the Falcons were called in to help coax more details from them. It didn't take long for Abu Ali to understand that the detainees knew something more crucial. These men had been part of an Islamic State cell that only a few days prior had bombed Mosul's bridges that span the Tigris River and act as the connective tissue for the city's commerce. The detainees told Abu Ali that this was the groundwork for a much larger offensive. They provided precise locations of camps in the Al-Jazeera desert to the west of Mosul that the terror group had been using to train fighters and store weapons. An attack, they told Abu Ali, was coming, the likes of which he couldn't imagine.

The information was reliable and the details chilling. Abu Ali passed his intelligence up the chain of command, saying all evidence pointed to an insurgent attack in early June. On May 31, during an urgent security meeting convened by the prime minister, the military commanders handpicked by al-Maliki to oversee Mosul dismissed Abu Ali's report as hysteria. When international officials in Baghdad reiterated the threat of an imminent attack on Mosul, which their own militaries saw as credible, the Iraqis told them that there was nothing to worry about.

The catastrophic consequences of writing off these warnings became clear on June 6, when Baghdad commanders started receiving panicky phone calls that convoys of pickup trucks and SUVs were rumbling across the northwestern Iraqi desert and firing artillery on Iraqi forces stationed around Mosul. In just two weeks, these terrorist shock troops had overpowered thousands of poorly trained and demoralized Iraqi security forces. The commanders leading the defense of the city—the same commanders who had dismissed Abu Ali's intelligence warnings—abandoned their positions on June 20, leaving mid-level officers in the field without supply lines of ammunition, food, or water. The Iraqi foot soldiers who had managed to survive the initial Islamic State bombardment were left to run for their lives.

The vanguard of insurgents couldn't believe their good fortune. They seized armored vehicles and tanks hurriedly abandoned by the Iraqi army and headed south toward Baghdad, killing thousands of soldiers and uprooting tens of thousands of families before arriving within ninety miles of the capital.

On the first day of the Muslim holy month of Ramadan, which that year started on June 24, Abu Bakr al-Baghdadi's spokesman announced the establishment of the Islamic State. Then, ten days later, on July 4, al-Baghdadi himself emerged in public. Speaking at Mosul's al-Nuri Grand Mosque, he promised to restore to his Sunni brethren their "dignity, might, rights and leadership."

In Baghdad, the al-Sudanis and the al-Kubaisis listened rapt, each in their respective living rooms, to the Islamic State declarations as they were rebroadcast by Arabic television news channels. Both families were aghast at what they heard. The fanatics were back in control. Thousands of Iraqi families were fleeing for their lives. Their poor nation—which had barely survived the first Al Qaeda onslaught a decade earlier—would be beset by another war with the jihadis.

In east Baghdad, Abu Harith and Um Harith were horrified at the nonstop pictures showing the terrorist fighters carrying out a wholesale slaughter of Shiite Iraqis, especially those in uniform who were training to serve in Iraqi security forces. Outside the al-Sudani home in Sadr City, mosque loudspeakers played dirges for the dead and implored all men to sign up to fight. Abu Harith feared that his sons would be deployed to the front line to help defend the nation from these devils.

Five miles away in Amariyah in west Baghdad, Professor al-Kubaisi, too, was immediately worried for his sons' safety. Both had high-paying, white-collar jobs, but that wouldn't protect them from retaliation fueled by the anti-Sunni screeds spewing from Shiite political parties and news channels. The scenes from Mosul and the communities devastated by the Islamic State were terrifying, with roads littered with Iraqi corpses and frightening tales of families who had managed to flee ahead of the terrorist wave.

Abrar was the only one who wanted to cheer the developments, but she was circumspect about airing her views in front of the family. Praising another Muslim's bad fortune would bring bad luck to the household, at least that's what her mother believed.

Online, however, Bint al-Iraq praised Allah for delivering her country from what she considered the Shiite government's yoke of oppression. Abrar had seen what the Iraqi armed forces had done in 2012 to innocent civilians, brave young men, families, and grandmothers, who had gathered peacefully in Ramadi and Mosul demanding civil rights and an end to corruption. Abrar and her online cohorts had predicted it would only be a matter of time before an uprising erupted. After all, what did the government expect would happen, after its blatant campaign to kill and silence its critics? Abrar hailed from a strong Sunni tribe in Anbar, where men did not back down against injustice. Real men, real Iraqis, would rather die than submit to oppression. And that is what the Islamic State forces promised the Sunnis of Iraq: liberation from the Shiite donkeys braying for the blood of honorable Iraqis.

When Abu Bakr al-Baghdadi repeated the demands of the Sunni protesters during his inaugural sermon as the caliph of the faithful in Mosul, it was as if God had answered her prayers. Here was a leader who wanted to restore dignity and meaning to the lives of Sunnis in Iraq. Thousands of Muslims worldwide were pledging their allegiance to this new ruler, and maybe this political eathquake would bring her the hope she had been searching for.

Even his rivals had a begrudging respect for Abu Ali's professional skills. When he identified his quarry, his concentration was intense. His observations were as sharp as a butcher's knife. His ability to see patterns from disparate data points was rare in the Iraqi government, as was his penchant for sitting quietly, deep in thought. Thanks to Prime Minister al-Maliki's patronage, the spymaster had freedom to work outside the confines of Iraq's lumbering bureaucracies, and had access to enough funds to build a network of agents and informants. Abu Ali's reputation among jihadis for sale was that he would make it worth an informant's while to share information, making him feel respected and even dignified in his decision to betray his terrorist comrades. Those skills mattered more than ever during the summer of the Islamic State's lightning strike across Iraq. The Iraqi army had collapsed and citizens had mobilized by the millions to defend their nation. Intelligence officers, including Abu Ali, were trying desperately to prevent Islamist zealots from taking over the country.

The month of July unfolded like a nightmare from which Iraqis couldn't awaken. An estimated four million people were trapped under the rule of the Islamic State. Hastily formed Iraqi militias and reconstituted military units managed to stop the advance of the terrorist army approximately ninety miles north of Baghdad. Abu Ali and his colleagues couldn't remember when they had last slept, as they rushed to a makeshift command center in Samarra and attempted to wring all information from their sources about the Islamic State.

Meanwhile, the country was rudderless. The backlash against al-Maliki that had emerged during the spring elections swelled in June. Now the entire nation was united in blaming him for the humiliating loss of one-third of the country's territory to the Islamic State, and leaving undefended a nation already traumatized by conflict. By August, with the country's future on the line, Iraq's international allies, including the United States and Iran as well as the Shiite religious authorities in Iraq, moved jointly to push al-Maliki out of office. In his place, they rallied around one of his party's stalwarts, Haider al-Abadi, a man educated in Great

Britain whose short stature and soft-spoken demeanor made him the mirror opposite of the man he replaced.

Al-Abadi's first day in office, September 8, 2014, was bittersweet. As he sat through briefings from his security chiefs and the newly formed U.S.-led coalition of forces that would be aiding Iraq in the military campaign against the Islamic State, the new Iraqi leader saw that they had precious little intelligence, let alone a plan to prevail. Al-Abadi remembered the dyspepsia he felt while reading through the dozens of pages of reports from the Iraqi intelligence services. The bureaucratese and obfuscating sentences couldn't disguise the fact that his commanders knew nothing about the capabilities and structure of the enemy that he didn't already know from reading the foreign press or turning on the television. The only exception was the quiet man who sat at the far end of the briefing table and away from the spotlight: Abu Ali al-Basri.

The Falcons' chief had compiled a briefing on the backgrounds of the Islamic State's known leaders, the majority of whom had Iraqi or U.S. military prison records. He summarized some of his unit's previous counterterrorism operations, the intelligence they had gleaned from the Mosul detainees in May, and his willingness and experience in working with the Americans. Abu Ali told al-Abadi that he didn't have the resources to prosecute a war against the heavily armed and capable enemy that Iraq faced, but he knew how to get eyes inside the enemy camp.

I don't know war, but I do know spies, he told the prime minister. We have several agents already feeding us information. But what I'd like is to get one of our own men inside.

One of the greatest assets a spymaster has is a network of trusted informants—human sources close to a target. Electronic surveillance can yield information about who attended a meeting or ordered an attack. But what eavesdropping can't determine is a sense of the enemy's morale, its commitment or intentions. Only well-placed human eyes or ears can do that, especially those who once pledged loyalty to the enemy, which is why such networks are so rare and valuable. The only thing better, in Abu Ali's view, would be a human source that he could plant himself, an undercover asset whose loyalty was unquestionably to Iraq.

Al-Abadi knew the inherent difficulties of recruiting spies inside the Islamic State, or inserting an Iraqi officer in the ranks of the extremist group. But the country's implacable war called for daring solutions. The new prime minister gave Abu Ali al-Basri more autonomy, a bigger budget, and the authorization to conduct independent offensive operations.

Abu Ali left the prime minister's conference room and walked down the grandiose marble-lined corridors and outside to the parking lot. He continued past the rows of shiny SUVs, past a small rose garden still blooming in the warm late-summer air, and on to the breeze-block building that had been the Falcons' home for six years. Men, he announced once he reached the cramped break room, we have a new task. We must infiltrate the Islamic State.

What was designed as a morale-boosting pep talk was instead met with barely disguised alarm. For months, his men had been monitoring the group's almost gleeful brutality, the torture and beheadings of hundreds of Iraqi police officers and soldiers that the group documented in daily propaganda videos. Abu Ali wanted a volunteer to pose as a jihadi, go undercover and report on the enemy's secrets. No one from the Falcons wanted to risk their life by going behind enemy lines, not even for their respected chief.

Later that week, Munaf and Harith al-Sudani laughed about what they saw as an outrageous request. Poor Abu Ali, Munaf remembers thinking. It will be a cold day in hell before that wish comes true.

Um Mustafa sat on the low stool worn smooth from years of kitchen work, in front of her a bucket of small pale-green zucchini. Holding a slender metal file, she flicked her wrist in small expert jabs to bore a hollow core, the technique honed from decades of cooking kousa mahshi, the Iraqi stuffed vegetable dish that was her daughter Abrar's favorite meal.

The al-Kubaisi family dined on the labor-intensive meal at least once a month, ever since Abrar had learned to walk, but Um Mustafa, a plump, smooth-skinned late-middle-aged woman with kind brown eyes, cooked it more often when she wanted to show one of her five children some extra love. That day, she reckoned, the al-Kubaisis needed as much comfort as her food could provide.

Her husband had come home unusually early in a torrent of angry shouting. Abrar, their witty and clever daughter, had caused a scandal that could irreparably damage her future.

Since Iraq had mobilized for war earlier that summer, all of Baghdad had been on edge. The killers from the Islamic State had enslaved two million Iraqis, and Baghdad's residents were tense with dread that a renewed wave of terror attacks would destroy their lives, a fear that Al Qaeda had instilled in them since the mid-2000s. But Abrar hadn't evinced much worry about that. Something else was consuming her, something that had changed her from the inside. Um

Mustafa had watched Abrar for weeks—her daughter was radiating an energy that propelled her from bed early in the morning and kept her up late at night. Whatever was going on, Abrar refused to acknowledge, let alone explain, what was percolating inside her so bright and fierce.

Um Mustafa was concerned, but she didn't know what to focus her anxiety on. On the outside nothing was different. Abrar hadn't bought any new clothes. She still wore the same demure black-and-white-patterned headscarves and shapeless cloaks that deflected attention like a respectable Muslim woman should. She was arriving home from work at the usual time, so her mother didn't believe she was conducting a secret love affair. But she felt with an instinct only a mother has that Abrar was harboring a secret of some sort. The only thing Um Mustafa could imagine was that her daughter might have fallen in love with someone at work, a development that would be a small miracle, as Abrar had never shown interest in men. Um Mustafa had always hoped that her daughter might get married and still enjoy her work, much like she had in her own life.

A phone call from the ministry administrator who supervised Abrar had deflated Um Mustafa's daydream. He had requested an urgent meeting with

Professor al-Kubaisi, and her husband soon discovered exactly what had been afoot.

There had been an argument in the hallway of the ministry, the director told him. It was unclear what had first sparked the dispute, but tempers had flared in an unseemly fashion. A group of women from Abrar's department quarreled about religion. There was shouting and cursing, and reports of Abrar hurling verbal bombs that could get her arrested. Several witnesses reported that she had insulted the country's iconic Shiite religious figures, and the prolonged row ended with Professor al-Kubaisi's petite and polite daughter slapping a headscarfed colleague's face.

Professor al-Kubaisi sat in stunned silence as he listened to the administrator recount the fracas. His daughter had not only embarrassed herself and her whole family, she had offended the department by wading publicly into the dangerous waters of sectarianism. That her language was similar to the jihadi acolytes of the Islamic State was frightening enough, but especially now, while the country was at war, she could be denounced not just as a bigot but as a possible terrorist sympathizer.

Abrar's boss, a Sunni like Professor al-Kubaisi, had survived the upending of Iraqi politics and the labor market that had occurred over the last decade, during which Iraqis favored in the Saddam era had been shoved aside for his victims, the nation's majority Shiites. He didn't want to lose his job now. He told Professor al-Kubaisi that there was pressure on him to fire Abrar, and while he respected Abrar's work, and the family's academic and social standing, he wouldn't abide the young woman's bigotry against Shiites and the risk it posed to his career. "Abrar is swimming in dangerous waters," the director told her father. "No one, not you nor I, can swim after her and survive."

Professor al-Kubaisi felt as if he had been caught in a freak sandstorm. He couldn't see a path clear of this catastrophe. Reputational risk was a virus that would infect the entire family, not just Abrar. There had to be a solution to mitigate the problem and save face for everyone, including the director. The two men drank coffee and reached a tentative agreement: Abrar would stop coming to work and the director would quietly push through a long-term leave of absence available to most ministry employees who wanted to pursue a continuing education. The al-Kubaisis would have to figure out a place for Abrar to hide away and study for another degree.

Professor al-Kubaisi returned home furious. His daughter had not breathed a word of what had happened. How dare Abrar put them all at risk with such an embarrassing situation. Her sharp words were certain to lodge in her mother's soul, like shrapnel from a bomb. She, after all, was Shiite, just like Abrar's colleagues. The family didn't practice the Shiite rituals, but Um Mustafa had raised the children to respect all the Islamic traditions. How had their daughter gotten so angry that she had forgotten her upbringing?

Um Mustafa listened to her husband's shouting without saying a word. Abrar sat on the living room sofa, withdrawn and unresponsive. When Professor al-Kubaisi's energy was spent, he announced he would go upstairs to lie down. Um Mustafa told her daughter to follow her into the kitchen where she could shut out the confusion of the outside world amid the clutter of familiar things: the colorful chrysanthemum-patterned plastic-coated tablecloth, the noisy tick of the wall clock, and the ingredients of her daughter's favorite dish.

For almost twenty-eight years she had spent each day with her daughter in this room, sipping tea and cooking meals. As a child, Abrar had attended religious studies classes taught by Um Mustafa in the elementary school just down the road from their home in Amariya. In the evenings, as a family, they watched television, and while they all agreed that the Shiite-majority governments elected after Saddam's overthrow had been a disaster for Iraq, Um Mustafa had never extended blanket criticism over the entire Shiite population. Somehow, however, her daughter had. Somewhere her anger had become so overwhelming that it had erased a lifetime of manners and mores and erupted in public. Um Mustafa didn't know what to say to Abrar, how to question her feelings or beliefs. They weren't a family that talked about such things. So instead of trying to find the right words, she kept her mind focused on preparing the food in front of her, hoping that the meal would communicate her love and concern.

Since that night, Abrar's family had devoted all their efforts to overturning the administrator's decision. Her uncle stepped in to petition for her reinstatement. Her father tried, too. All the old-timers at the ministry, men who were Sunni like her family, pushed as much as possible on her behalf. But they no longer held much power. Abrar's outburst had made her too poisonous to touch.

While they were busy trying to repair the damage she had caused, Abrar was looking to the future.

Since the Islamic State had declared its caliphate in northern Iraq, Abrar had known that her life would change in a radical way. The reemergence of an insurgency to take on the corrupt Iraqi government and sweep it away was just what she and her online friends had been eagerly awaiting for years. For months, these confidants had been pressuring her to join them in the Islamic State. But it wasn't easy for a young woman in Baghdad, who wasn't married and had never spent a night of her life away from her parents' home, to figure out a way to cross the front lines to reach the territory under the group's control. Losing her job turned out to be one of the best things that could have happened, as it gave Abrar the opportunity to act.

It wasn't in Abrar's nature to do everything she was told, but she had no problem following the directions of the mentor she had found online in her favorite online chat room, Shumukh al-Islam. Abrar had followed Abu Nabil's discussions online for years. Unlike her father and her uncle, Abu Nabil didn't have a university degree. But he was a proud Iraqi, born in Samarra—the same city as the Islamic State leader Abu Bakr al-Baghdadi. His education came from the battlefields of Iraq after the Americans invaded in 2003. He fought the occupiers and was even imprisoned in the notorious Abu Ghraib prison. He proudly wore the scars of torture he said he received from the crusader forces. He never tired of saying that God willed him to survive his detention at their hands, so he had dedicated his life to jihad, a calling he learned about in prison.

Abu Nabil quoted Koranic verses beautifully and he always had an answer for the problems that she posed online about the injustice the government had shown to Sunnis like them, about the nefarious plans that Western powers had to keep Iraq weak and divided. He wasn't boastful, but it was plain to Abrar that Abu Nabil was a man trusted by the leaders of the Islamic State, especially after they promoted him to oversee the province around Samarra, a job that made him one of the twenty most important leaders of the group.

Over time, Abrar developed intense feelings for him. She had been warned over and over not to trust anyone, but she told Abu Nabil about the secret research she had devoted herself to in the often trying circumstances that she faced in her work in Baghdad. She even told him her real name. In return, Abu Nabil encouraged her laboratory research, and became one of her most enthusiastic cheerleaders. Those infidels don't appreciate your talents, he told Abrar repeatedly. The caliphate needs the best and the brightest Muslims to join our cause. We need people like you.

Which is how Abrar came to be waiting, in the darkness just before dawn, for her brother's friend Akeel.

Her parents thought he was going to drive her to the airport, help her pass through the layers and layers of time-consuming checkpoints, and see her off for the plane to Turkey. Um Mustafa had been awake since four A.M. to feed her daughter breakfast. She sipped tea while Abrar paced from the sofa to the window. Her daughter looked excited, as any young woman would be for her first trip abroad. Um Mustafa had no reason to doubt what her daughter had told her: that she had been accepted to study at a cancer research institute in Sakarya, Turkey, a university city between Istanbul and Ankara, and that the institute would accept her transfer for graduate studies even though the semester had already started. They were living in strange times, and Um Mustafa assumed that the war with the Islamic State had disrupted life in Turkey as well. In fact, she was simply happy that her clever girl would have a chance, at age twenty-eight, to finish her studies, since God had thrown up so many roadblocks for her in Iraq.

Abrar's version of reality, though, was different. She had made a deal with Akeel: in return for a payment equivalent to her monthly salary at the ministry, he would drive her from Baghdad to his hometown of Al Qaim, a town on the Iraqi border with Syria that since the summer of 2014 had become a major operations hub for the Islamic State. Akeel's brother-in-law was the Islamic State commander in charge of military operations there, and Abrar planned to petition him for an introduction to the senior leadership of the group. Inside her suitcase, she was bringing a gift for the caliph, the fruits of the project that she had secretly worked on since she was an undergraduate at Baghdad University, first at its labs and then at home in a small, unused back room of her parents' house.

The previous night, she had carefully sealed her sticky creation into empty powdered milk canisters, which she then hid amid her clothes in her baggage near the front door. No one would bother checking them, she reassured herself. No one ever paid attention to someone like her, a quiet woman in a headscarf. Still, she had been warned by her online friends to be careful.

Akeel pulled up in front of the al-Kubaisi house just after the muezzin called the dawn prayer at the mosque down the street. Um Mustafa invited him inside for some tea. She had promised herself that she would not cry when it was time for Abrar to leave, but she couldn't help herself. Her daughter was so jittery that she pulled away from her mother's embrace. Mama, we have no time, Abrar told her. I can't miss my plane. I'll contact you once I reach Turkey.

Ten hours and a hundred miles later, Abrar and Akeel drove along a potholed dirt road out of Iraqi-controlled territory across a smuggling route that Islamic State guards and Iraqi frontline soldiers had erected off the highway outside Ramadi in the western desert. Akeel had prepared the paperwork for them to cross from Iraqi territory into the land of the caliphate. The sleepy solider on duty let them pass without a glance. For the first time in her life, Abrar gave thanks for the corrupt ways of Iraq's Shiite rulers. She had reached the caliphate. Al Qaim was just a straight shot ahead.

Abrar saw little of the wan, dusty border town after arriving at Akeel's family home, especially after he told his brother-in-law what she had smuggled in her luggage. The dark, sticky paste was a toxin called ricin, a chemical warfare agent banned due to the lethality for those who inhale it. Abrar told the local Islamic State military council that she could help them use this powerful weapon to smite their enemies. She had learned from her mentor Abu Nabil that the terror group had its own chemical weapons department. Her gift, she hoped, would allow her a place among their ranks.

The commander, Akeel's relative, dutifully sent a report to his superiors in Mosul, telling them about the unusual Iraqi woman scientist. Abrar, however, would not receive permission to travel there until the Al Qaim commander had approval from the Iraqi leadership, meaning that Abrar had to settle in for an indefinite wait.

Her patience soon wore thin. News of the biological weapon had spread through the ranks of militants in Al Qaim and several commanders had requested demonstrations. The problem was, not a single one of the men had a high school diploma, let alone a basic knowledge of chemistry, genetics, and proteins. Abrar felt like she was an elementary school teacher as she walked them over and over again through the same explanation as she tested the ricin first on rabbits, then on a captured street dog.

Ricin kills because it prevents your cells from manufacturing proteins, Abrar said. It breaks your body down from the inside, she went on, using the simplest language that she could.

She directed them to the internet, where stories abounded about how ricin was used in covert operations. She told them how the Soviets used it to murder an anti-government dissident in the 1970s and how American terrorists used it to attack a U.S. congressman.

The bearded men seemed to understand better as they watched the animals fall ill and stop breathing. But they insisted on seeing its effect on humans as well. They brought her a prisoner, a Syrian farmer who lived just across the narrow ribbon of river that, until the caliphate, was the natural border between the two countries. Abrar forced him to swallow some ricin, but he didn't die immediately. He fell into a coma, which worried Abrar because she thought she had perfected her dosage levels to induce death within minutes. The jihadis didn't seem to care. They seemed fascinated watching the farmer's slow march toward death, how his organs failed and his skin turned blue as his body struggled for oxygen. They shot him at that point, to make sure he had actually expired.

The worst part of her stay in Al Qaim was the lack of her own computer or phone, which were forbidden to all women in the Islamic State. Abrar couldn't abide being cut off from communications with everyone, including Abu Nabil. She hoped that he would have prepared an introduction for her to the chemical weapons scientists in Mosul. But apparently the Islamic State's bureaucratic wheels moved as slowly as Baghdad's, no matter how many times a day one prayed.

While she waited on word from Mosul, Abrar was confined to Akeel's family home. Like the other women in town, she had to wear the head-to-toe chador that the Islamic State insisted upon for modesty. Women in the caliphate weren't allowed to work, and the tribesmen who controlled the border had never heard of a female scientist before, especially someone like Abrar who came from a well-known Iraqi family. She suspected that they felt sorry for her. After all, she wasn't married, and she had no children, though she was almost thirty years old. They had dedicated their lives to fighting the Iraqi government, but they couldn't understand why a woman might make the same decision.

She willed herself to be patient. As soon as the permission from Mosul arrived, she would travel there and go to the city's fabled university where the Islamic State had converted laboratories for its weapons and scientific research. Working in a hub of activity along with other ideologically driven scientists like herself was her dream. There, they would revive the spirit of Islam's Golden Age, when Muslim scientists and scholars revolutionized all understanding of the natural world.

Unbeknownst to Abrar, foreign intelligence agencies had already taken a keen interest in the Islamic State's activities inside Mosul University. Even before Abu Bakr al-Baghdadi had declared the establishment of his caliphate, the plans for building a weapons research and development program had been put into motion and the group was well on its way to producing a chemical weapon. Until they captured Mosul, what the Islamic

State militants lacked was a safe and secure environment to build such unconventional bombs.

The laboratories at the university were some of the most advanced in the country. They had stocks of anthrax that researchers from the agricultural science department used to create vaccines for livestock. The chemistry and biology departments stored low-level enriched uranium to conduct medical testing as well as precursor chemicals that could be mixed to build crude chemical weapons.

Iraq is one of the few countries in the world with firsthand experience of the horrors of chemical war, first on the battlefields of the Iran-Iraq War in the 1980s and then when Saddam attacked his own citizens in 1988 in what his victims call the Anfal genocide. Memories of these horrors remain vivid for many of the country's new leaders, including Abu Ali, so when the Islamic State took control of Saddam Hussein's largest chemical weapons production and storage facility in July 2014, Iraqi security officials were nervous, if not downright hysterical.

Rumors that the terror group had stockpiles of chemical weapons spread like wildfire in September 2014, when Iraqi troops deployed in Salahuddin province were rushed to the hospital after a heated battle. They arrived with strange symptoms, with many vomiting and struggling to breathe. Doctors concluded they had been exposed to chlorine gas used by Islamic State fighters. Abrar's mentor, Abu Nabil, was the top militant in the area at the time of the presumed chemical weapons attack.

For the next several months, Western and Iraqi intelligence officials, including the Falcons' cyber unit, had picked up chatter from online encrypted social media sites where pious Muslim scientists like Abrar congregated. Many were bragging that they were joining the Islamic State to help build an arsenal of banned weapons. These devotees exchanged formulae and recipes for manufacturing anthrax, sarin, and mustard gas. Western intelligence agencies kept close watch on the main routes that foreign Muslims used to smuggle themselves from Turkey into the lands controlled by the caliphate for any sign of shipments of raw materials that could be combined to produce banned chemical weapons. In one such case, in 2015, the Americans informed the Iraqis of a three-person cell that had managed to bring a container truck filled with lab equipment and large industrial exhaust fans into the country. The Falcons monitored and then arrested the men, who finally admitted under interrogation that they had been operating under orders from commanders in Mosul.

The terror group's propaganda machine sought to capitalize on the fear unleashed by the reports of its chemical weapons. In Baghdad, the news prompted panicked speeches in Parliament and endless discussions on evening talk shows about how families could protect themselves in such an attack. Iraq's defense ministry ordered an emergency delivery of a thousand gas masks and gear for frontline soldiers. To try to preserve morale in the nation, Prime Minister Haider al-Abadi placed a blanket ban on any reports of biological or chemical attacks.

And still, Abrar waited in Al Qaim. In mid-June 2015, a month after she had arrived, she received the word she had been waiting for. She would be welcome in Mosul. All she needed to do was find a way to get there. The Americans had targeted four senior Islamic State commanders in three different airstrikes on the roads surrounding Mosul, and the militants had stopped all movement of its top officials around their Iraqi capital.

As it turned out, the road for Abrar to Mosul wasn't straight and narrow. It was long, winding, and went through Baghdad.

That's how the Islamic State ordered her to travel. Abrar would need to go home, they said, get a passport and gather as much money as she could, and then follow the well-established pilgrimage route that foreign Muslims were using to enter the eastern territories controlled by the Islamic State. It wasn't a woman's place to question the command. So, for the first time in three weeks, Abrar was provided access to a phone. She called her father and said she would be back in the Iraqi capital the next day.

Abrar's homecoming was slightly strained, her mother recalled, but no one in the family had any reason to doubt her story. She told her parents that she needed to return to Iraq for transcripts and other paperwork. She also said that she had started a part-time job at a pharmaceutical lab in Turkey that was owned by a Syrian man, because the cost of living and studying was simply too high. While back home, Abrar renewed her passport and emptied her bank account of nearly $10,000, which she had saved from her salary at the ministry. Professor al-Kubaisi was worried despite Abrar telling him everything was fine. He called the family friend whose daughter was also studying in Turkey. Everything seemed to check out. As Abrar had always been strong-willed, they found no reason to prevent her from leaving home again.

On July 23, for the second time in as many months, the al-Kubaisis said farewell to their daughter, and this time Abrar really did fly to Turkey. Once in Istanbul, she did what hundreds of foreign Muslims had done already that year. She slipped onto a bus headed to Gaziantep along the Syrian border and enlisted smugglers to take her to Raqqa, the Islamic State's capital in that country.

To her surprise, the welcome she received there was cool, bordering on hostile. For the first time in years, Abrar wondered if she had made a mistake. Instead of the respect and understanding she had enjoyed from her fellow Iraqi allies, the Syrian contingent of the Islamic State was haughty and suspicious. She presented her permissions and invitations from some of the most respected members of the caliphate, Abu Nabil and the Al Qaim commander. But that meant nothing to the Syrians. She was assigned a guard from the hisbah, the Islamic State's morality police, and sent to one of the group's dormitories for women, where she was kept isolated for forty-five days, cut off from the internet and her phone.

Abrar never knew why she had been confined, whether she was suspected of being a foreign spy, or whether the Syrian section of the caliphate was simply more inefficient than the Iraqi side. If her guards thought they would break her, they were mistaken. She liked to be alone, left with her notebooks and research, with time to review her scientific work and to think of new ways to manufacture potent toxins.

Finally, her hisbah guard came to inform her that she would be driven to Mosul the next day. The vehicle, a white minivan similar to those that ferried commuters around Baghdad, was full of women passengers. It was a strange phenomenon, driving down a roadway with strangers, each hidden in a flowing head-to-toe black niqab. They had all been in the caliphate long enough to know the rules. No talking, lest their voices tempt a male warrior away from his work. No flesh could be revealed for the same reason. So hour after hour, they sat in the back of the van, hands covered with thick black gloves and roasting like chickens in the early summer heat. Unlike Abrar, the rest of the women were going to Iraq as brides for the fighters there. Despite the code of silence, the women in the back of the bus with Abrar whispered amongst themselves. Some prayed, some gossiped. Some prodded her, wanting to know what Abrar was doing in the caliphate. Abrar remembered what Abu Nabil had always told her: trust no one. She had no idea who the people were under all the layers of clothing. Perhaps this was some sort of final test before she reached Mosul. She didn't utter a word, knowing that her journey was almost at an end.

On September 10, the temperature had passed 100 degrees before ten A.M. when Abrar's bus rumbled into Mosul. She was immediately escorted to the home of an Iraqi widow and her two teenage daughters whose jihadi father had been killed the previous summer. There, Abrar unpacked and prepared for another long wait.

It felt strange to be in a world that was so familiar yet so different. Everyone who came and went from the house spoke in the familiar tones of Iraqi Arabic, but the women acted like they had grown up on another planet, not in the country that Abrar knew. Um Sarah, the widow, had been married at the age of fourteen and had barely finished elementary school. Her husband had been a well-known Al Qaeda member for years, and she lived an extremely sheltered life as a result. The only people she knew in this city of three million were her own family members. Although the caliphate boasted that Mosul's streets were safe and the people happy, Um Sarah refused to go out unless she was accompanied by a male guardian. She had lived in the neighborhood for two decades and her husband was a known martyr for the cause, but she still worried that the roving bands of religious police would harass or beat her. Even inside the house, where she belonged, she obeyed the strictures of Islamic State teachings. She forced her daughters and Abrar to keep their hair covered, even though no man was around to see them. She didn't have a phone, let alone a television or computer, in the house. The only news from the outside world came when Um Sarah's brother arrived with food and groceries.

Growing up in Baghdad, Abrar thought of herself as religious, but she had no qualms about going around town whether on foot or in a taxi driven by an unrelated man. She saw nothing wrong with a woman studying science or participating in religious discussions online, with a group of strangers. God had given her a brain and she considered it his will that she use it.

But in Mosul, she felt limitations at every turn. When word came that the chief of the Islamic State's chemical weapons program would meet her, Abrar had expected to be taken to the laboratories at the university where she knew that the scientists worked. She thought she would have a chance to discuss her research in the professional environment of a research facility, where people wore white lab coats and spoke her academic language. Instead, the man who introduced himself as the head of the Islamic State's chemical weapons program, an imposing Iraqi man named Abu Ruwaydah, came to Um Sarah's house. The commander sat in the family majlis on a deep cushion on the floor and Um Sarah's brother then erected a wooden screen in the middle of the room and only then invited Abrar into the room. Abu Ruwaydah was a true Salafi jihadi. He refused to look at a woman who was not a relative and insisted that Um Sarah's brother serve him tea to keep the strict policies of gender segregation espoused by the Islamic State.

So it was that after all her preparation and anticipation, Abrar found herself dressed from head to toe in a black polyester robe, in a full face veil, talking through a screen to the man who would decide whether or not to accept her into the ranks of the caliphate's weapons research team. Rather than discussing the glories of scientific achievement, the only thing Abu Ruwaydah was interested in was the maximum number of deaths a ricin attack could cause.

Sister, Abu Ruwaydah kept saying, God wills us to vanquish the unbelievers and purify this land. Science will be a weapon that we wield for this purpose and this purpose only.

Abrar was unsure about how to present her ideas. Clearly, it would be an uphill battle to make this man respect anything she had to say. The silence in the room was broken with a telephone ringtone, a Koranic prayer favored by the Islamic State. Abu Ruwaydah started speaking on the other side of the partition. Clearly, the restrictions against communication devices didn't apply to him. As the call droned on, it became obvious to Abrar that he didn't have any interest in her.

When silence settled again, Abrar gave Abu Ruwaydah the same tutorial that had eventually won over the men in Al Qaim. She told him how a natural plant indigenous to Iraq could be harvested and purified to toxic levels and how an injection or concentrated spray could kill scores of people at once.

The commander was unimpressed. Sister, God commands that we kill hundreds, not dozens. What can we do to cause mass destruction? That's what we want to know, he told her.

Abrar asked if she could speak to the other scientists, see what materials and compounds they had access to. Only then, she said, could she make a report about what chemical weapons could be built.

In reply, all she heard was a deep grunt. Abu Ruwaydah had apparently stood up from the floor cushions that lined the wall on the other side of the partition. She didn't know whether that was his way of rejecting her request, or just the sound that a man of his stature made when he was ready to leave a room.

Two weeks later, she had an answer, though not the one she had hoped for. She was ordered back to

Al Qaim to help the weapons production facility there. She would not be allowed into the university research labs after all.

Abrar traveled back to Raqqa and then to Turkey—her dream shattered. When she crossed the border back into Gaziantep, it was already late November. The autumn winds had a chilly bite and the summer clothes she had packed in Baghdad were no longer adequate. But Abrar didn't remember feeling the cold. As soon as she arrived in the Turkish city, she rushed to an internet cafe, the first online access she'd had since leaving for the caliphate almost three months earlier. She logged in to her old chat rooms, desperate for word of her online friends. Mainly, she wanted to ask Abu Nabil, her mentor, who had guided her to this point, what she should do next.

Her heart sank when she saw no new messages from her mentor. Indeed, there was no sign of him online at all. None of his aliases had been active on any of the regular social media discussion forums for weeks. Abrar scanned the Islamic State media outlets and even went to Twitter to see if there was any word about him. He couldn't have totally disappeared, she thought.

And then she saw the post, a florid eulogy for a man described as the Lion of Jihad. Abu Nabil had been killed on November 15, in a U.S. airstrike in Libya. The Islamic State's eulogy described how the group had sent the Iraqi commander to the North African country to establish a branch there. Before he left, he hadn't even bothered to say goodbye.

Abrar spent the next few days alone in her hotel room in Gaziantep praying for guidance. She didn't want to go back to her old life in Baghdad under the rule of the corrupt government. If she returned to Al Qaim, she would find herself saddled with all kinds of restrictions. She might be able to conduct her own research alone, but the tribesmen from that part of Iraq didn't respect a female scientist. Surely, in time they would force her to marry and have children, like a dutiful wife of the Islamic State.

The only option left, she thought, was to show the world what she was capable of. She would plan her own attack. It would take months of organizing, Abrar realized. She would need more raw materials, and a lab, items now harder for her to access in Baghdad after losing her job.

Abrar's email in-box was full of messages from her brothers, anxious for her to send updates so they could assure her parents that she was safe. One morning as she walked from her hotel to the internet cafe intending to finally answer their entreaties, Abrar noticed an ad from a Turkish pharmaceutical company seeking workers with a scientific background. It was written in Arabic—apparently aimed at the thousands of well-educated Syrians who had flooded the Turkish city to escape from the war in their homeland.

Abrar wrote a brief message back home. Tell Mama and Baba I am well. I have a job at a pharmaceutical plant and am studying hard.

Chapter 14
War Hits Home

The first light in Sadr City was a gloriously soft, peachy orange, its warmth diffusing through morning air heavy with dust and humidity. It was Thursday, August 13, and Um Harith hadn't slept well on her makeshift rooftop pallet in Baghdad's stifling heat. No one could remember when the temperature last dropped below 100 degrees, or the last time it had rained. The nation had been at war with the Islamic State for a year, but in the summer of 2015, it wasn't the war on terror that most upset Iraqis in Baghdad. Top of their complaints was the heat, of which there was too much, and electricity, of which there was too little. Only the wealthy could get a good night's sleep under the whirr of ceiling fans or generator-fueled air-conditioner units. Most of the capital's four million residents woke up sweaty and cranky.

In Sadr City, people had more reasons to complain than most. It wasn't just that summer temperatures were unbearable. An estimated 30 percent of the district's young men were on the front lines, either with the regular Iraqi forces or the Shiite militias mustered to fight against the Sunni extremists. And while the sons of Sadr City were being killed to liberate Iraq, their families struggled to make ends meet.

The open-air Jamila market, just a couple of miles from the al-Sudanis' home, was a mainstay for Sadr City residents living paycheck to paycheck or waiting for their monthly government welfare stipend. Unlike in the modern supermarkets that dotted Baghdad's more fashionable neighborhoods, they could haggle and buy in bulk. And since Jamila market opened early, just after the first morning prayer, shoppers could get home well before the heat of the day.

That thought spurred Um Harith to roll over on her sleeping mat and stand up slowly, testing her arthritic knees. She said a small prayer of thanksgiving as she padded her way past her sleeping grandchildren to the staircase. Along the way, she poked the oldest, Harith's boy Muamal. Her job as the al-Sudani matriarch was to make morning tea. His job, as the oldest unmarried boy in the house, was to run to the market to buy the bread everyone would eat for breakfast.

Um Harith later remembered how quiet and still the streets were that morning. She had opened all the doors and windows hoping to catch some breeze to help cool the ground floor where all sixteen people in the house would soon gather for breakfast. She heard her daughters-in-law cooing at babies as they changed diapers, and her granddaughters up on the roof bickering as they folded sheets. Soon a couple of the toddlers straggled into the kitchen where she was putting teacups and sugar on the breakfast trays.

Um Harith was pouring boiling water over the thick carpet of tea leaves in the kettle when she felt a tremor under her feet. The girls tidying up on the roof felt a blast of air hit them like a punch. They turned westward, in the direction of the shock wave, and saw a massive plume of black smoke spiraling into the sky about a mile away.

Baghdad residents of a certain age, who had lived through the insurgency of the mid-2000s, have acquired an unusual, macabre life skill: a pitch-perfect ear for identifying explosions, whether the bone-crunching thud of a vehicle bomb, the whistling of incoming mortars, or the icy crackle of a concrete building imploding from the concentrated fury of a suicide bomber. For that reason, the al-Sudani women didn't need to turn on the television or radio to know that a truck bomb had detonated somewhere in Sadr City.

They were right. Earlier that morning, while the night sky was still a deep bruised black, a clean-shaven man dressed in a white dishdasha drove through the capital into the Shiite district of Jamila, passing several security checkpoints without any soldier bothering to check what he was ferrying. The owners of the small shops that framed the market arrived at daybreak to open their corrugated metal shutters and green iron doors, and didn't notice the stranger or his white truck, which resembled dozens of vehicles that appear at the market, filled with vegetables to sell. Regulars at the market told the police that they had never seen the truck or the driver before that morning.

The bazaar started filling with shoppers after the muezzin's call to prayer, and the air filled with the crisp aroma of freshly washed parsley and fried cumin. At around six A.M., the driver started hawking his wares, calling out in a singsong voice about his cheap and tasty tomatoes. When a crowd lined up around him, the driver triggered his explosives. The blast, about half the size of the Oklahoma City truck bomb, ripped through the throng of people, flinging bodies to rooftops a street away and severed limbs even farther.

At the al-Sudani home, Harith's mother offered a silent prayer for the women soon to be mourning their loved ones killed in the blast. Her head was still thick from sleep, but the clamor of a half dozen toddlers pleading for their breakfast pulled her from her thoughts. She suddenly realized that Muamal wasn't home with their bread. Dear God, she thought, I sent him to the market.

Her heart pounded and she screamed, sending a fresh shock wave through the house. Where's Muamal?

The grandchildren who weren't already crying started to, shaken by their grandmother's outburst. Three of Harith's brothers came running down the stairs, stirred by the panic in their mother's voice. Raghad, Harith's wife, was right behind them.

I sent him to Jamila, Um Harith repeated, her voice rising with fear. Raghad bit on her fist to keep from crying out. Although every nerve in her body was twitching, urging her to action, she couldn't run out on the street to look for her son. It was unseemly for a married woman to be seen in public, especially in her house robe. No matter the reason.

Instead Raghad scanned the assembled family. Neither Harith nor Munaf were home. Raghad assumed that her husband had spent the night on duty at the Falcons headquarters, something that had become routine since his recruitment a year earlier. In their absence, she locked eyes with Munthir, the youngest of the al-Sudani siblings, who had been the last to come downstairs, stretching his slender arms above his head, like he didn't have a care in the world.

The twenty-year-old moved through life with the smug optimism of a youth who didn't know discomfort. In a house full of women, Munthir, with his warm brown eyes and aquiline nose, always had someone to bring him tea, cook his meals, and wash his clothes. His father's parenting style had mellowed considerably by the time the baby of the family arrived. Everyone agreed he was spoiled, but his sense of entitlement had never ripened to arrogance, and his good nature always helped lighten the mood. Raghad, for one, was glad Uncle Munthir liked to be pampered at home. It meant he was around for Muamal when his father wasn't. Both Um Harith and Raghad rushed to the young man, and Munthir ran out of the house without having to be told twice.

As the nightingale flies, Jamila market is ten minutes from the al-Sudani house. By foot, the journey is almost twice as long. That morning, Munthir ran faster than ever before, down the narrow residential lanes, past the haunted house and the car repair shop owned by his father's friends. Each time he passed a familiar face, he shouted the same question. Have you seen Muamal?

Smoke billowed in oily black plumes and the stench of fuel oil hung heavy in the air. Closer to the market, roads filled with an army of anxious men who, like the al-Sudanis, had relatives at the market that morning. The screams of the wounded merged with the sirens of emergency vehicles.

Munthir slipped past the cordon of police trying to keep hysterical family members from the devastated market street, his slender frame lost in the muddle of bodies. Moving down the street, he thought he had stepped into hell as he tried to maneuver around the thick puddles of blood and severed limbs. He walked by charred corpses and groups of men with blood caked on their cheeks and matted in their beards. Munthir couldn't tell the difference between offal for sale at butchers' stalls and the human body parts strewn across the narrow street.

After just a few moments in the devastation and chaos, he knew that it would be impossible to find Muamal here. He needed to call Harith and Munaf and tell them what had happened. Where Munthir stood, the oily thick smoke of burning plastic and fuel oil tasted like the end of everything.

It had been a year since Prime Minister al-Abadi's election, but Abu Ali and his Falcons had made scant headway in penetrating the top echelon of the Islamic State. In fact, the entire war, which was being waged with the help of three thousand American troops, a thousand additional international forces, and the reconstituted Iraqi armed forces, was at a stalemate, according to General Ray Odierno, the U.S. Army chief of staff and former head of U.S. forces in Iraq.

If the existential threat posed by the terror group wasn't dire enough, al-Abadi's staff was also grappling with a looming financial disaster. They were struggling to pay government salaries month after month, let alone procure the ammunition, weapons, and food that the Iraqi army needed to march to battle each day. Iraq was on its knees and the prime minister was desperate for good news.

Abu Ali and the Falcons, meanwhile, had only gained a few yards in the marathon to build their network of spies behind enemy lines, people who could reliably tell them what the Islamic State was planning and how and when it would strike next. Abu Ali was still convinced he needed a man inside the militant group to make headway. But he didn't have it in him to order any of his men to what would be an impossible mission—he needed a volunteer.

When Harith's phone rang that Thursday morning, he and Munaf were sleeping off what had become a normal night of work for the Falcons team. They had worked until the early hours scanning internet chat groups. The brothers were dozing in the Falcons' headquarters, with the window shades pulled down against the summer sun. The hallways outside the office space were empty, as the morning shift had not yet arrived. Munaf and Harith had slept through the familiar sounds of the call to prayer, so it was the punchy voice of popular Lebanese singer Nancy that had finally pierced the fog of sleep. It was Munthir on the line.

Brother, where are you? Have you heard the news?

No, Munthir, what's happened? Is it our parents? Are they okay?

Harith, turn on the television. Jamila has been bombed. The streets are flowing with blood.

Ya Khara, Harith said. Shit.

He quickly sat up on the sofa, spilling paper cups of cold tea and candy wrappers onto the floor.

Munthir, tell me what has happened. I can't find a remote. What's going on?

Brother, forget that. Listen to me. I'm at the market, surrounded by bodies. No one can find your son. Do you hear me, Harith? Come home. Now! Muamal is missing.

What the fuck are you saying? Where is Muamal? What has his mother done with him? What is going on? Harith kicked Munaf to wake him up.

He passed the phone to his brother as he scrambled for the remote.

Munaf later recalled straining to hear his younger brother amid a cacophony of ambulance sirens in the background. Munthir's voice coming through the handset was higher and thinner than normal and he kept repeating himself.

Brother, find out where the injured are being taken. It's chaos here. We need to find Muamal. He's missing.

Munaf looked over at Harith, his face blue from the refracted glow of the television. On the screen was a live report from Sadr City showing the familiar low-slung buildings at Jamila market. Walls had been sliced apart like ripe melons from the blast, disgorging burned bodies and wrecked vehicles. Men struggled to move unwieldy, blackened bundles from the street. Munaf could see from the strips of tattered jeans and colorful cotton that those unrecognizable bundles were once human.

Harith's face, already haggard from a lack of sleep, turned ashen. His hazel eyes were slightly unfocused. He seemed dazed.

Harith, move! Let's go. We need to get to Sadr City. We need to go find your son.

Munaf drove with Harith back to Sadr City as quickly as possible in the morning rush-hour traffic. He maneuvered his car around minibuses bringing students and medical staff to the teaching hospital and sped past the lines of cars in the right lanes heading to work at the Oil Ministry. The overpass spanning the dried-up canal between Baghdad and Sadr City was jammed with emergency vehicles and black pickup trucks painted with the insignia of the Jaish al-Mahdi, Sadr City's local militia.

Munaf's black Samsung cell phone had become hot to the touch from his nonstop calls. He and Harith were desperate for news about casualties, but the chaos meant that there was no central clearinghouse of information. So Munaf did what all Iraqis instinctively do: use his wasta. He dialed all of his contacts from the police academy and those colleagues called up for emergency duty in Sadr City that morning, asking all of them whether they knew of any teenage boys who had been taken to the district's emergency rooms or to the morgue.

In the hour it took Munaf to drive from the Interior Ministry to the Jamila market, he had learned nothing of importance. No lists of wounded had been compiled. All available emergency workers had been called to duty, but their priority was to save lives, not identify the dead.

Munaf pulled his white Hyundai sedan up on a curb near the market. Frustrated, he handed his phone to Harith, who had remained unusually quiet in the passenger seat. His eyes were dull and he had completely withdrawn. Munaf hadn't seen his brother so despondent since the day the family had learned he had flunked out of university.

Harith, pull yourself together. Call Munthir, Munaf told him. Maybe he has some updates. I'll be back in ten minutes.

Munaf jumped out of the car and started jogging toward the bomb site. The smoke-filled air was now choked with the smell of burned flesh. His police ID in his hand, he pushed past the groups of weeping men. He passed three ambulances and peeked inside each, seeing bloodstained sheets. The medical crews were on their third roundtrip from the hospital to the blast scene. They suggested he head to the Imam Ali Hospital. Lists of dead were being collected there.

The carnage was worse than he imagined. As he returned to the car, he wasn't sure what he would tell his brother. When he arrived, he saw Harith drumming his fingers on the windowsill. It was the most animated he had been the whole morning.

Munaf, drive home. Munthir told us to come right away, Harith yelled. I don't know anything else because your phone died.

Munaf made a U-turn and tried to navigate the sedan home. He cursed, his left arm thrusting impatiently out the window, trying to direct the crush of police and crowds out of his way. Years of wearing a uniform had given him a feeling of invincibility. But that morning uncertainty and doubt had washed away his confidence, leaving him vulnerable and unsettled.

When he pulled into the rutted lane in front of the family home, Harith jumped out of the car even before Munaf had managed to turn off the ignition. He ran through the gate and went inside. Seconds later, Munaf caught up to him in the kitchen. To his relief, there stood Muamal, wrapped in a bear hug by his father, tears streaming down both of their faces. Raghad and the other women agitated around them, like hummingbirds.

Muamal's clothes were streaked with soot, but he had no injuries. Um Harith stood in the corner of the kitchen next to the electric stove with her hands in the air, thanking God that her grandchild was safe. Munaf asked Muamal where he had been, his voice more gruff than he intended. The boy pulled his head away from his father's chest, his eyes red, and he struggled to answer.

Without warning, Harith stepped back from his son and cuffed him upside the head.

Answer us, son. Where the hell have you been? We shouldn't be thanking God, we should be cursing you. Look at what you have done to your grandmother!

Muamal's eyes welled up again. Not from the slap, Munaf thought, but from the still raw memories of the bloodshed he had witnessed that day.

The boy told his family that he was already halfway home from the market when the bomb went off. He didn't know why, but he felt an urge to go back to see what had happened. So he walked back by the bakery where he had been just a few moments earlier. In place of the bustling business were the burning corpses of customers. Blood was all around him.

The boy broke down. He couldn't finish his sentence.

That was enough for his grandmother to intervene. Enough already, said Um Harith. Everyone needs to drink their tea and just calm down.

Muamal joined the grown-ups in the family room, each settling in their familiar place around the sofra.

The family's single television set, mounted on the corner of the wall, was turned to the Sharqiya news channel, which was broadcasting nonstop the gore and carnage along the main streets of Jamila market.

By evening, sixty-seven people had been confirmed dead, twenty-eight of whom, like Muamal, had been children under the age of fifteen.

Abu Ali al-Basri spent the afternoon in the marble-lined hallways of the prime minister's office, walking between meetings with irate political leaders demanding what the Sadr City residents wanted: a stop to the attacks. He felt the quiet anger experienced by counterterrorism officials the world over. Politicians were never interested in the grinding daily work needed to successfully stave off attacks—they were only interested in venting their frustration when a mistake occurred. Not that Abu Ali could blame them for their rage, given the bloodbath at the market.

The war was in a critical stage. Prime Minister al-Abadi needed morale to stay as high as possible, but photos of children's bodies covered in white shrouds in Sadr City weren't helping. Men who had crowned him as Iraq's leader were now publicly trying to tarnish him.

Harith al-Sudani as a toddler with his parents, Abu Harith and Um Harith

Harith al-Sudani with his father and two cousins at an amusement park

Harith and Munaf al-Sudani playing soccer with relatives

Harith and Munaf al-Sudani with a parrot, posing as colleagues while working for the Falcons

Harith al-Sudani on his first visit to the sea, during his vacation to Lebanon

Harith al-Sudani's funeral poster commemorating his death in the line of duty for Iraq

Raghad al-Sudani and two of her children in 2020 in the al-Sudani family home

Raghad al-Sudani on the Islamic pilgrimage to Mecca in 2019

Abu Harith, Raghad, and her children

Abu Harith and Um Harith in the al-Sudani family home

Abrar al-Kubaisi, taken after her detention in 2016 by the Falcons

Harith al-Sudani and colleagues from the Falcons Intelligence Unit

Prime Minister al-Abadi felt like he was living through a live version of *Groundhog Day*, a film he had seen years ago in London. He hadn't understood many of the jokes at the time. But after he became leader of Iraq, he had a new appreciation for the dark humor of finding himself chairing the same meetings and same security problems over and over again.

No one in the world, not the American military advisers nor the Iraqi security commanders al-Abadi had appointed a year earlier, could give the Iraqi leader the data he needed: the numbers of hard-core militants who were trying to destroy his country and kill his citizens; the cells that threatened the capital; realistic assessments about the threat to the nation's critical infrastructure and its venerated religious sites. Al-Abadi read report after report on the subject, both classified assessments based on electronic surveillance carried out by the American-led coalition fighting the Islamic State as well as analysis published by so-called experts in Washington and London. But the Iraqi leader was left with an unshakable conviction: no one really knew what they were talking about.

Abu Ali understood al-Abadi's frustrations. In many respects, he shared them. His men excelled in producing granular detail about specific Islamic State cells, and in the course of the past year, the Falcons and the Iraqi security forces had gained a much clearer idea about the relative strength of the terror group. But these successes didn't amount to much when a single man could penetrate the defenses that the military and police had erected around the capital, drive a sophisticated truck bomb into one of the most influential Shiite neighborhoods in Baghdad, and kill family members of soldiers risking their lives on the front lines. In political terms the Jamila market attack was critical for al-Abadi, because it opened him up to charges of incompetence. It also risked tearing open the sectarian wounds that had started these terror attacks in the first place.

Since Abu Ali had taken the job as head of security for the new prime minister, he had participated in enough training sessions and after-hours conversations with his foreign colleagues to understand the philosophical battle raging among intelligence professionals since 9/11. Was there a high-tech solution to the war on terror, like the Americans believed? Could tapping enough phone communications and emails, or photographing enough vehicles and buildings from satellites thousands of miles in the sky actually stop the enemy? The volume of data the Americans produced was enormous, and Abu Ali saw how obsessed his colleagues were with PowerPoint displays composed of colorful charts, graphics, and measurable results. He looked with envy at the ways these wily bureaucrats used their information to justify and expand their operating budgets.

But Abu Ali wasn't waging a global war. He needed to discover specific details of specific terror attacks, the movements of one man with one bomb driving toward a market teeming with farmers and housewives, or where a top Islamic State commander might be next week. Because of that, the spy chief found himself on the side of the intelligence debate that emphasized the need for human sources, not big data and high-tech wizardry.

Abu Ali knew from his own years in the Iraqi underground and from the search for Al Qaeda in Iraq that while it was important to know who was planning to meet whom, something that a drone or phone GPS locator could track, the best-quality intelligence came from knowing what people inside those rooms or sitting on a park bench were saying. Having an agent there, inside those rooms and being part of those conversations, was the gold standard he wanted to achieve.

The interminable meetings in the Green Zone that Thursday after the bombing kept Abu Ali at work well after midnight. He made his way to his white Toyota Land Cruiser and waiting driver to head home for a couple of hours of sleep before resuming work. As they cruised down Baghdad's empty streets, past boulevards of locked shops and homes where families were praying for their children's safety, he wondered again how he was going to find someone who might walk into the lion's den.

In Sadr City that afternoon, a thicket of funeral tents blossomed like bitter weeds. The al-Sudani house was unusually quiet, as everyone absorbed how lucky the family had been to cheat death. Harith's siblings returned from the mosque with news that their neighbors had not been so fortunate. At least six people the family knew had been killed that day.

In the tradition of the neighborhood, mourning was a drawn-out public event, done in full view of family and friends. The rituals drew crowds of relatives and people tied by intricate webs of social and professional relationships, political party affiliation, or support for the same sports team. Harith had never been an enthusiastic participant in these events. Invariably, such large gatherings meant to console the grieving led to gossip about the living. Harith may have secured a prestigious job with the Falcons, but in a tight-knit community like Sadr City, no one had forgotten his past shame.

That is why it took the al-Sudanis by surprise when Harith announced he would pay his respects to the bomb victims, especially the families who had lost children in the blast.

For three days, he went from tent to tent, becoming increasingly morose at the photos of child-size shrouds and pine coffins he saw. Between reciting prayers and drinking cups of bitter coffee, Harith was pulled aside by groups of neighbors, their eyes hardened in anger. They wanted to know the identity of the bomber, who had failed to stop him, and what officers like Harith were doing now to keep them safe.

Information was the one balm that might bring solace, but he had nothing to tell them. The Falcons didn't know who organized the attack, let alone who carried it out, and this filled Harith with a renewed shame.

Days later, after the mourning tents had been disassembled, Harith woke up in a cold sweat. He had dreamed that Muamal was trying to escape from a burning building and he watched helplessly as his son tripped and fell, his body engulfed by the flames.

He wanted to burn away this ache of failure and find some way to assuage his neighbors' pain. It was time once again, Harith felt, to step up.

Chapter 15
Volunteering for Danger

A week after the Jamila market bombing, Harith came to a decision. As he rode to work with Munaf, he told his brother that returning to his old work routine would be unbearable. Sitting in an office pecking for clues online about the whereabouts and plans of Iraqi jihadis was no longer enough. He said he wanted to do more to stop the terrorist threat ripping their communities apart.

When Munaf pulled into the Falcons' compound, he watched Harith jump out of the car and walk across the gravel-lined driveway to the one-story bungalow where the Falcons' senior officers worked. Once inside, Harith told Abu Ali's assistant that he needed an immediate meeting with the spymaster.

At first, the aide was reluctant to interrupt the boss. He'll want to see me, Harith told him. Tell him I'm ready to volunteer.

The words made the adjutant jump. He rushed into the inner office where Abu Ali was behind his polished mahogany desk, speaking on the phone. He passed the chief a handwritten note and watched Abu Ali's eyebrows perk up as he read it. His boss stood up slowly from his black leather desk chair and put his phone to his chest to prevent the caller from hearing his order.

Call the commanders and schedule an urgent meeting for this afternoon, he told his aide. Make sure Harith and Munaf are there as well.

With that, Abu Ali called the elder al-Sudani brother in to see him. An aide brought them glasses of mango juice and steaming black tea. Harith, dressed in his dark blue uniform, sat in the hard-backed gilded chair facing Abu Ali's desk. He didn't touch the drinks. He looked the spymaster in the eye and said what Abu Ali had been waiting months for someone to say.

It's my duty to God and to our nation to prevent the deaths of any more children, Harith declared.

Abu Ali came around from his desk to sit next to Harith. He let the weight of the words settle in the silence. As he did so, their differences in age and rank receded and an intimacy bloomed, cultivated in the realization that both men were taking a step of enormous consequence. Why now, my son? What has changed for you? Abu Ali asked.

Harith paused, then answered. It's my own son, he said. I almost lost him. I didn't realize until this week how much I have failed him as a father. By taking on this mission, I have the power to save him from at least one terrible fate.

Later in the day, Abu Ali convened eight senior intelligence officials cleared to hear about his undercover operation. They didn't know what to expect before the gathering, so when the Falcons' chief described what he had in mind and introduced the al-Sudani brothers, several gasped in shock. Others expressed reservations about embarking on such a mission. It's too dangerous, said one. It's suicide, said another. We can't sacrifice any more good men, said an officer from the mukhabarat. His agency had considered a similar operation but hadn't found anyone willing to carry it out.

As the officials around the table offered up their views, Harith stood resolute and impassive. Munaf, however, looked deeply pained. He knew from their conversation in the car that morning that his brother had been pondering some sort of drastic action, but only now did the gravity of Harith's decision truly sink in.

Listening to the other commanders at the table, Munaf silently agreed with their assessments. The risk was enormous. Yet he was torn. He certainly knew, better than they did, how clever his brother was. He also knew that if anyone could succeed, it would be Harith, who was consumed by his quest for an achievement to erase the shame he had brought on the family and to make their father proud of him at last. At the same time, he also understood that if the Falcons successfully infiltrated Harith into an Islamic State cell, then his brother was unlikely to make it out alive.

The officials, having said their piece, turned to Abu Ali and waited for him to weigh in. The moment of decision had arrived. All morning long the Falcons' chief had vacillated between the selfish joy of having the answer to the problem that had vexed Iraq for so long, and the guilt of sending one of his men on a mission whose odds of success and survival were so small.

Abu Ali took a deep breath and looked directly at Harith. The spymaster didn't know much about the younger man's childhood, but he knew that anyone who could survive and thrive in Sadr City was a person with stamina. But was it enough to withstand the psychological and emotional rigors of an undercover assignment among terrorists, when just a single slip could expose him and end his life? Could Harith, an Iraqi Shiite, trick Sunni militants into thinking he was one of them? Could Abu Ali live with himself if his officer was killed?

There was no time to mull over his misgivings. Most of them were imponderable anyway. Harith was his man, and thus his responsibility. Abu Ali would have the final say.

Putting his own emotions aside, the spy chief declared Operation Lion's Den underway. Harith walked out of the room and grabbed his brother into a big bear hug.

Finding a volunteer for the mission was only the start of this complex plan. The next hurdle was to train Harith, a man born and raised in the most well-known Shiite neighborhood of Baghdad, to pass as a Sunni militant. To do that he needed a detailed cover story, a new accent, and more psychological tricks to survive the strain and the isolation of the long, dangerous mission.

For five weeks, Harith trained in seclusion. He learned how to pray like a jihadi, talk like a jihadi, and fake that he was a jihadi.

Abu Ali was pleased with Harith's progress, like that of a young bird learning to fly. Like all Iraqis, Harith had grown up with a prodigious capacity for memorization. Schools demanded children learn lessons by rote. As a boy Harith had memorized the entire Koran and volumes of poetry. Those mental powers would be applied to remembering details of meetings he attended and the exact words of conversations that he overheard while embedded in the Islamic State cell.

Abu Ali's plan was to have Harith impersonate a native of Iraq's Anbar Province, a hotbed of extremism. His young officer learned how to flatten his accent to mimic the region's diction. The long nights he had spent online posing as an Islamic State member in chat rooms had made their language and rituals familiar. But there was something else about Harith that Abu Ali had noticed, a trait that made him an even more valuable asset. He had an uncanny ability to compartmentalize, keep the kernel of feeling and care for his wife and children deep inside of himself, along with the wound that he had been carrying for so many years. Abu Ali had no idea, until the psychologist brought in to evaluate Harith's mental preparedness told him, that he had the shame of his family wound so tightly inside of him. If he had the strength to detach himself from those deep-set emotions, Harith would be able to detach himself from the traumas and pressures of being a spy.

When the training was complete, Abu Ali reconvened the eight intelligence officers briefed on Harith's mission for a chance to test his skills. The Falcons officer rattled off the biography of the jihadi whose identity he was adopting, then the names and biographies of the Islamic State men whom his alias would have known. The officer who trained Harith on the cover story said he was satisfied. The psychologist signed off on Harith's mental fitness, as did the tactician who tutored Harith about dead drops and other ways to communicate when he was undercover.

Abu Ali then turned to Munaf, who was also sitting in with the examiners. He knew his brother better than the other men in the room. The spymaster wanted to hear him say that Harith was ready to embark on the mission before he gave his own green light. The younger al-Sudani knew that his brother was determined to seize this chance, to prove himself once and for all. The fact that the mission was dangerous wouldn't dissuade him. Munaf just hoped that Harith's need to show their father had been wrong wouldn't jeopardize the mission.

Sir, I believe Lieutenant al-Sudani is ready, Munaf told the assembled staff officers.

Then, I'm satisfied as well, Abu Ali replied. Let's go.

Munaf buckled his seat belt and watched his brother lean forward for a better view out the window. Their Iraqi Airways plane was starting its descent into Beirut and Harith was craning his head to glimpse the shimmering turquoise sea below them.

Imagine a life where you could see that every day, he said to Munaf, a hint of yearning in his voice. The sea could make anything more beautiful. Even Sadr City.

It was September 2015 and Abu Ali had given the al-Sudani brothers a week of leave as an acknowledgment of how difficult the training had been. They had a chance to go anywhere they'd like, at Abu Ali's expense. Neither brother had ever left Iraq. Neither had ever been on a plane. They agreed straightaway to choose the Lebanese capital for their vacation.

Beirut has long been called the Paris of the Middle East—a city of exciting nightlife and big dreams. Unlike Cairo, which also has a reputation for casinos and cabarets, the Lebanese capital also has a large population of Shiites, and for two young men from Sadr City that fact added an extra layer of comfort. They would be visiting a new country, where they could let off some steam and try new things, but also where life wouldn't feel totally unfamiliar.

Abu Ali hadn't spelled this out when he presented them with this gift, but Munaf realized that their boss had wanted Harith to have a chance to clear his head and experience new surroundings. Abu Ali rarely spoke of his years living in exile—though on occasion he mentioned fondly the bracing steel gray waves of the North Sea and the towering green trees of Sweden. Munaf knew that his boss loved Iraq, but also had a deep appreciation for his experience abroad.

Maybe, Abu Ali told Munaf later, Harith would see something unique, something that captivated him so much that he would reconsider his decision to enter the lion's den. Abu Ali, like Munaf, had misgivings about the operation. Not about Harith's readiness, but because they were both averse to risk. It had been drilled over and over into Munaf's head that there would be no easy way to extricate Harith if he needed help. If something went wrong, his brother would be easy prey.

Munaf planned their weeklong vacation, and he meant to do all they could to have a good time. On the two-hour flight from Baghdad, the brothers drew up a wish list. They wanted to visit Al Omari Mosque, a place where the great Islamic warrior Salahuddin had declared his victory over the crusaders. They also wanted to try whiskey—which was what all the rich men in Egyptian films drank—and maybe even the famous Lebanese anise liquor arak, which Munaf had been told about countless times since his years in the police academy. Women, his fellow cadets had assured him, love to drink arak.

As the plane touched down in Beirut, Harith added a new objective. He wanted to put his feet in the soothing waves of the sea.

Their six days were magnificent. They wandered through Beirut's noisy, crowded streets until the early hours of the morning. They spent a small fortune entertaining themselves at a cabaret. They left the nightclub unaccompanied and too drunk on cheap whiskey to care about anything but reaching their hotel room beds. They took reams of photos of the city's iconic sites, pictures that Munaf still keeps on his phone. In one shot, Harith stands next to the lighthouse on Beirut's corniche, the seafront walkway, wearing a red shirt and a half smile. It is hard to tell from the faraway look in his eye whether he appreciated at that moment just how happy he was.

On their last night in the Lebanese capital, Munaf and Harith sat at a seafront cafe smoking shisha and picking through a bowl of pistachios. Harith had wanted to spend his last hours of vacation watching the surf. The blaring songs of a Lebanese pop star drowned out the hiss and roar of the waves, but the clean briny smell filtered up from the sea as a pleasant reminder of their trip.

Munaf saw that his brother was more at peace than he could ever remember. Harith had never managed to actually swim in the sea—neither brother had learned how to swim—but nevertheless the rhythm of the tides and the blue water had touched something deep inside him.

Harith exhaled the tobacco smoke from his water pipe, leaned back in his chair, and started chatting about the future.

Inshallah, he told his brother, Baghdad could be rebuilt as beautifully as Beirut once our war with the Islamic State is over.

When that happens, he added, maybe we could change jobs. Find a way to get the canal along the perimeter of Sadr City filled with water and open our own cafe, one like this.

Munaf had made it a point not to talk about work that week. He was under orders from Abu Ali to help his brother relax. Still, the spy chief had told him that there would be a moment during the brothers' trip when talk of a different life, a different future, would arise. When that occurred, Munaf should test Harith's resolve.

If he expresses any doubt, any wavering, we will stop this plan before it ever gets off the ground, Abu Ali said.

Munaf could hear the elation in his brother's voice as he savored the fantasy of running his own cafe. It was perhaps the only time he had heard Harith speak frankly about his dreams. This was the moment that Abu Ali had foreshadowed.

As you wish, brother, Munaf replied. As you wish.

But why wait for that dream to come true? If that is really what you want to do, I'm sure Abu Ali would help you settle down here. Forget Iraq and the shit overwhelming our nation. Learn how to run a cafe here in Beirut and then come home when the war is finished. Only a few people know about the undercover mission. No one would blame you if you decided to pull out.

Harith immediately sat up straight, startled by Munaf's words.

Brother, you misunderstood. I am ready to sacrifice my soul for Iraq. I only meant that once we succeed in this, then maybe we could do something else. Besides, he added, more pensively, What would our father say if he ever found out I quit?

The brothers' arrival back in Sadr City raised a flurry of excitement at home. Nismah, Munaf's wife, had spent a whole morning agonizing over what to wear. Raghad, Harith's wife, didn't bother. She knew from past experience that whatever effort she put into her appearance, she wouldn't hold Harith's attention for very long.

However, as the family gathered to hear about the brothers' adventures and sample the Lebanese sweets they had brought home as gifts, Raghad sensed that something had changed in her husband.

Instead of retreating with Munthir and his other brothers to the main family living room, Harith invited her and the children to sit together in their own apartment upstairs. She made tea while he sat with their three children on the floor cushions in the largest of their two rooms, which doubled as the children's bedroom and the living room.

Usually, Harith was slightly impatient with the children, if not downright gruff. That day, though, the children squirmed under his unexpected attention. He asked each of them about their favorite subjects in school and the cartoons they watched on television. He tried coaxing Muamal to talk with him about soccer, which position he liked to play and which of Baghdad's professional teams he supported. By the end of the afternoon, when Raghad went back downstairs to help her mother-in-law clean up, the atmosphere wasn't exactly cozy. But it was better than the flat silence Harith normally brought home with him.

Although they had been married more than a decade, Harith had never confided in Raghad. That night, as they shared the same bed, she had no inkling that her husband was embarking on one of the most dangerous espionage missions ever imagined.

The next morning started like an ordinary day. Harith and Munaf dressed for work. They ate breakfast with their parents on the sofra, Harith taking his usual two pieces of samoon, the chewy oval-shaped Iraqi bread. He drank his tea and went out to the car without any sweeping expressions of affection to the family. Munaf was the only one who knew that, if everything went according to plan that day, the family wouldn't be seeing Harith for a long time.

The two drove through Baghdad's morning rush hour in silence. There were no last-minute doubts, nor teary paeans to his wife or children. For Harith there was no looking back. His upcoming mission would start his road to redemption. Along the clogged highway, Harith began his transformation. He asked Munaf to turn off the car stereo. True jihadis didn't listen to music. He emptied his pockets of gum wrappers and any other piece of paper that would hint at his real identity. Finally, the long vowels that characterized the accent spoken in the Shiite south and their descendants in Sadr City disappeared.

By the time Munaf had pulled into the Falcons' offices in the airport complex, Harith was well on his way to becoming Abu Suhaib, a disgruntled, brash Sunni laborer from the Baghdad neighborhood of Adhamiyah.

Chapter 16
Launching the Mission

By the fall of 2015, the Falcons had built a detailed chart of the Islamic State's sympathizers and cells in and around Baghdad. Their plan to infiltrate Harith into the group hinged on one of Abu Ali's informers, a low-level Islamic State bag man who had been arrested while distributing cash to the terrorists' supporters in the Iraqi capital.

Mohammed al-Jibouri was a petty criminal with little going for him before the Islamic State took him in. He hailed from a prestigious Sunni tribe, but his branch of the family was a rotten limb. When the Falcons arrested him, two of his brothers were already in prison, one for larceny and the other for terrorism. Al-Jibouri wasn't working for the Islamic State for ideological reasons. He was in it for the money.

Abu Ali capitalized on al-Jibouri's greed. He told his prisoner that if he made himself useful to the Falcons, then they could make life easier for his brothers in jail. And, if al-Jibouri agreed to work as a mole, Abu Ali would pay him $200 a month, a better salary than he had received from the Islamic State. The offer was good enough for al-Jibouri to accept. Soon, he was spilling important information about the terrorist safe house network in Baghdad.

His success as an informer prompted Abu Ali to return him to the streets of Baghdad that summer. The spymaster wanted him to resume his former life with the Islamic State, but this time as a mole for the Falcons. He assigned Munaf al-Sudani to be al-Jibouri's handler, thinking it was a good way for his bright, young officer to get more field experience. The rules of the agent-running game were simple. Al-Jibouri would get paid if he checked in and provided quality information. If al-Jibouri failed, his brothers in jail would bear the consequences.

While the rules were straightforward, the implementation was not, as Munaf learned when he drove his new spy from jail to Baghdad's central bus station.

Get yourself reestablished with your cell, Munaf told him, and then contact me in ten days to make your first report.

The younger al-Sudani brother watched the double agent walk away into the sharp sunlight and lose himself in the throngs of commuters waiting for the bus to take them home. When Munaf drove away, he prayed that he hadn't just let a cold-blooded killer back onto the streets of Baghdad.

Abu Ali and senior officers in the Falcons had warned Munaf that running agents, especially his first, would require patience, like waiting for the birth of your first child. The lull between meetings conjured up worst-case scenarios. But there is nothing an agent runner can do to change or speed up events.

While Munaf waited for first contact, his already heavy shisha-smoking habit got worse. He lost his appetite and couldn't sleep. On the morning al-Jibouri was supposed to report in, Munaf took his phone with him everywhere, even to the bathroom, so that he wouldn't miss a message. But the day came and went with no word.

Munaf wallowed in misery, thinking he had failed without having had a chance to properly start.

But on the eleventh day since disappearing at the bus station, al-Jibouri sent the agreed-upon signal, and information started flowing. His old comrades believed the cover story that Abu Ali had provided him to explain away his long absence. He explained that he had been kidnapped by rival tribesmen, a common enough occurrence in Iraq that his Islamic State handlers didn't question him. Instead, they gave al-Jibouri $9,000 to hand out to Islamic State sympathizers around Baghdad. He told Munaf that he had picked up exactly where he had left off before his arrest.

Over several weeks, Munaf coaxed from al-Jibouri vital information that fleshed out the Iraqi government's knowledge of the terror group. The agent relayed the identity of the man responsible for planning bomb attacks in the Iraqi capital: Abu Qaswarah. They learned that the logistical route used to ferry explosives to Baghdad started in Al Qaim—the border town with Syria where the Islamic State had a factory for building suicide vests and vehicle bombs. The supply line stretched through Anbar Province and to the small town of Tarmiya, an hour's drive from Baghdad.

For all this crucial information, Abu Ali wanted al-Jibouri to play an even more important role. He wanted to insert Harith within this network so they could foil the Islamic State's planned attacks, not merely understand how they were organized. Al-Jibouri, he thought, would be the tool to place his officer inside.

The spymaster and Munaf instructed their mole just how to accomplish this. He would tell the Islamic State commanders in Qaim and Mosul about a new trusted recruit from Baghdad named Abu Suhaib who wanted to help their cause and bring down the Iraqi government. Abu Suhaib—Harith—stood ready to pledge allegiance to Abu Bakr al-Baghdadi and serve the caliphate in any way needed.

Once al-Jibouri had set the lure, it didn't take long for the insurgents to snap at the bait. Abu Qaswarah contacted Abu Suhaib on Telegram, the encrypted social media platform that the jihadis used to communicate. He told the would-be Islamic State recruit where they could meet and when.

Travel to Tarmiya and attend Friday prayers, the Islamic State commander said. You will find what you are seeking there.

On a normal day, without much traffic, the trip from Baghdad to Tarmiya takes about forty-five minutes. The morning that Munaf drove Harith to his rendezvous, cars zipped through the highway checkpoints as families traveled northward toward Ramadi for the weekend. Munaf remembered the trip lasted all of a heartbeat.

Abu Ali had decided that the brothers would maintain a light footprint on this trip. They had no way to do a proper reconnaissance before they arrived. They didn't know the layout of the mosque, the number of exits, or who within the congregants were Islamic State members. Knowing its history as a popular Al Qaeda recruiting area, Abu Ali thought the numbers of enemy sympathizers would be extensive.

Munaf told his brother some of the same things he had told his first agent. He reminded him of the protocols to alert the Falcons of an immediate threat, or request a briefing. He counseled patience in the quest to get the terrorists to trust him.

They pulled into an empty farming road a kilometer or so away from the mosque. The prospect of entering a room full of violent extremists didn't seem to worry Harith. Munaf was far from certain that was the proper attitude to have.

He went over the details one more time. Munaf would wait for two hours for Harith to return. Don't be cocky. Don't be arrogant. Don't forget that they are going to watch you like hawks.

Don't worry, brother. I can handle myself, Harith replied. He looked at his watch, took it off his wrist, and handed it to his brother. Whatever happens, don't come looking for me. I can handle myself. With that, he got out of the car and walked toward the mosque.

Munaf sat in the car with the windows rolled down, cursing himself for not bringing anything to eat or drink. He couldn't leave his post for something as trivial as a bottle of water. He would just chalk that up as a lesson for future stakeouts.

Soon, in the distance, he heard the familiar sound of the muezzin calling the faithful to prayer, and then the voice of an imam over a loudspeaker reciting the Koranic verses for the weekly prayer. Munaf decided that he would start counting down the two hours to Harith's return only after the worship service was complete. The jihadis likely had agents inside the mosque monitoring Harith during the prayer, and that the meeting with the Islamic State cell would really only start after the imam had bid everyone to go in peace.

Munaf's instincts were right. But as the afternoon shadows got longer, there was still no sign of Harith. He tried not to let dark thoughts overwhelm him. His phone was filling up with messages from the Falcons' headquarters, his fellow officers seeking updates. Munaf replied over and over with a simple phrase "nothing to report." It was almost time for early evening prayer when Munaf finally saw a flash of movement about three hundred yards down the road.

To Munaf's relief, the figure approaching wasn't a stranger. The jaunty step of the approaching man was his brother's. His smile, which was large enough to light up the darkening sky, broadcast his success even before he reached the car. I'm in, he told Munaf. I'm in.

Chapter 17
Inside the Lion's Den

Harith woke up feeling lost, the scent of the sea in his nose and his hand pressed over his mouth in an unconscious attempt to stifle a scream. Sprawled on a thin foam mattress, he was completely disoriented until he realized that a nightmare had jolted him awake. The same terrible dream that had been plaguing him for months.

The scene, as he later described it to Munaf, always started peacefully. He saw himself, like a character in a film, walking along the seashore in Beirut, the waves lapping at his knees. Through the glare of the sun he could see an auburn-haired woman in the distance, someone he thought might be Nisreen, his first love. He felt a surge of joy, and then, without warning, a giant wave crashed over his head and swept his legs out from under him. His body tumbled underwater and a riptide dragged him far out to sea. Saltwater poured down his throat and Harith was overwhelmed with the certainty that he was going to die.

That vision stayed lodged like a fishhook in Harith's stomach, as he opened his eyes and remembered where he was. It was December 2015 and he was surrounded by sleeping Islamic State terrorists at a farmhouse in Tarmiya—his home for three months now.

Since Munaf had dropped him off for his first meeting with an Islamic State commander, Harith had worked to make himself—or rather Abu Suhaib—indispensable at this hub of the Islamic State juggernaut. His introduction to the terrorist group, via Mohammed al-Jibouri, had worked like a charm, and his training in cultivating the right mix of obsequiousness and radicalism accomplished what he and Munaf had hoped. Just a few short weeks into his undercover mission, he had become a protégé of the commander of his terrorist cell, a fifty-four-year-old farmer named Abu Mariam, who had been fighting with Al Qaeda and then the Islamic State since 2005.

The rectangular room where Harith and his comrades were sleeping had become as familiar to him as the back of his hand. Three sides were lined with twin-size foam mattresses covered in an embroidered

red fabric. Against the fourth wall, nearest to the door, stood two cupboards made of cheap, dark brown lacquered wood. One held glass teacups, a sealed canister filled with sugar, and aluminum serving trays. The other cupboard was filled with thick, furry polyester blankets that the men used as bedding at night. A ceiling fan hung in the center of the room, one of the few luxuries at the farm to be used when Abu Mariam had enough fuel to operate the farm's generator.

The daily routine at the farm seldom varied. At sunrise, the men would wake, put away their blankets, and pray. Then, after a quick breakfast, Abu Mariam would turn the room into his schoolhouse and teach the cell members the ABCs of terrorism. The men learned how to wire explosives and conduct surveillance. In the evening, they learned the religious texts beloved by Abu Bakr al-Baghdadi. In between, the men would help with the farm chores and do calisthenics. Abu Mariam urged his men to keep alert. Their commanders in Mosul could order them on a new operation at any time.

For the first three weeks, Harith was watched every waking minute, and never was left alone. While he was outside doing his chores, his new Islamic State comrades searched his shoulder bag, in which he stored his extra set of clothes. He felt that every word he spoke was scrutinized. But thanks to his Falcons training, Harith slowly won the group over. He proved himself adept at the technical lessons and Koranic study. He gave them no reason to doubt his dedication to the cause. Furthermore, he offered a valuable attribute to the cell, part of his undercover persona created by Abu Ali al-Basri. Abu Suhaib had an Iraqi government identification card that showed him to be a native of Baghdad, and a car that was registered in Baghdad. This meant that Abu Suhaib could travel in and out of the capital more easily than the other men in the cell, and certainly more easily than Iraqis whose ID showed they hailed from territory under control of the Islamic State.

A little more than a month after he joined the cell, Abu Mariam walked over to Harith after the evening call to prayer and handed him a cell phone. It turned out to be proof that Harith had passed the initial tests set out by the commander. On the line was an Iraqi Islamic State commander in Mosul, Abu Qaswarah, the man in charge of attacking Baghdad. He told Abu Suhaib that they had chosen him to be their angel of death, the man who would drive the terror group's explosives to the Iraqi capital. With no more preamble, Abu Qaswarah then ordered him to drive two grooms, the group's euphemism for suicide bombers, to Baghdad.

When he handed the phone back to Abu Mariam, Harith uttered a prayer of thanksgiving. The cell leader must have thought his pupil was overwhelmed with the honor of the mission he had been assigned.

My son, may your acts and deeds bring glory to God, he told Harith.

The call from Mosul elevated Harith's status at the farmhouse. In the two days that followed, he settled into a steady, if uneasy routine, as he and Abu Mariam waited for the arrival of the grooms. Meanwhile, Abu Suhaib was excused from all farm chores so he could prepare for this offensive operation.

The evening before his recurring nightmare, just before sunset, two skinny young men appeared on the dirt road leading to the farm. The grooms, one a Tunisian and one an Iraqi, were barely old enough to shave. The two newcomers barely said a word that night. When it was time for bed, they slept soundly, to Harith's amazement. Instead, the undercover officer lay wide awake, feeling the crush of pressure about the ways in which his mission could go wrong.

In the morning, he would either be an accomplice in the deaths of dozens of civilians in the capital, or his Falcons comrades would prevent the attack by intercepting him and his passengers. It was the first big test of Abu Ali al-Basri's goal. The Falcons were supposed to neutralize the threat, but also make it appear to the Islamic State leaders that Abu Suhaib had been successful by pumping out misinformation about a supposed bomb attack. In Baghdad, while he was training, the logic seemed elegant and easy to Harith. But in the dead of night in Tarmiya, surrounded by hard-core terrorist devotees, the task appeared daunting, if not impossible. The one chance Harith had to succeed was to alert the Falcons that he had been activated. At the moment, the one tool he had to communicate with his unit, a basic Nokia phone, was hidden in his car, far from his room in the farmhouse.

Between his bed and the door was his commander, Abu Mariam, and four other jihadis. Harith knew from many nights of studying all possible ways out of the farmhouse that there was no way he could sneak out of the room without waking the others.

And if Harith somehow managed to reach his phone, or perhaps even escape from the farm? He could walk forever through the countryside and not get anywhere. He knew no one who lived in the vast plain between him and Baghdad. Outside of this farmhouse, he was a stranger to many and an enemy to all.

Alone in the dark with his fear, Harith wrestled with the urge to flee. He finally conceded that he had few options and many formidable adversaries, starting with

Abu Mariam. He hadn't told Munaf yet, but he was in awe of the Islamic State commander. Abu Mariam was lean like a corn stalk, with a surprising, wiry strength in his long limbs, which, like his face, were darkened from years of working in Iraq's punishing weather. Upon meeting him, jihadis immediately noticed that the leader did not have a single gray hair, although he was as old as many of their fathers. In their eyes, this was a sign of fortitude, a victory over the hardships he had endured as a farmer and the stress of fighting the American military for more than a decade. Like Harith's father, Abu Mariam was a man of his word.

For all his admiration, Harith had no doubt about how Abu Mariam would react to treachery. In his first week at the farmhouse, while he was still being tested by the cell, Abu Mariam had followed Harith to the outhouse behind the main building. Harith's true identity was a deeply held secret, but he felt like Abu Mariam's eyes had drilled straight into his soul. If you ever betray us, I will be the one to cut your throat, he said.

Outside the farmhouse, wild dogs barked in a choppy chorus. Inside, Harith needed something to distract himself from his unreachable phone, so he focused on the two grooms sleeping less than ten feet away.

In a few hours, when the sun broke the horizon, Harith and these two strangers would wash and ritually purify themselves, and then kneel side by side to beseech God to answer their prayers. Harith would ask to be kept safe. He had no idea what would go through the minds of the young men. Like him, would they think of their mothers? Would they dwell on their own best memories of their brief lives?

In the darkness of the Tarmiya farmhouse Harith tried to calm his anxiety by recalling his own best moments, a device that he learned in his training. He recalled how, as a child, his mother would lay her soft, plump hands on his forehead when he was ill. How Nisreen's lips moved as she recited his poetry. But recrimination and regret intruded. During their vacation, he and Munaf hadn't found the time to drive north out of the capital to see Lebanon's snowcapped mountains. Would he ever have another chance to do that? Would he have been happier if he had walked away from his life in Iraq and instead stayed abroad?

Later, as his mission wore on, he revealed to Munaf his ultimate fear: not just that he would die, but that he would die with regrets.

As he waited for the sky to lighten, Harith turned to the only psychological balm that he knew. He quietly recited the most comforting prayer in the Koran, the same one that the two men he was driving to Baghdad would recite later in the day before they tried to kill themselves.

Soon, a rooster sounded the start of the day and Harith's attention shifted. The man lying nearest to the door yawned. In the corner, Abu Mariam started coughing, the hacking noise of a habitual smoker, as he stood and moved to each mattress to shake each man awake.

Harith lay on his back and stretched, stealing glances at the two grooms. They looked so young in the early-morning light, Harith thought. They also looked well rested.

He resisted the temptation to stare at them as they dressed, each one buttoning the freshly pressed shirts that had been hanging from hooks on the back of the door. The clothes were the uniform of a Baghdad civil servant, chosen by Harith to help them blend in with the army of rush-hour commuters who drove to Baghdad each day.

Abu Mariam started the call to prayer, a signal that the men should pick up their sleeping pallets and place their prayer mats down on the floor. The grooms were given the places of honor closest to the commander, giving Harith a chance to observe them from behind. He detected no nerves from either of them, although this was likely the last day they would spend on Earth.

Once prayers were finished, several of the men went outside with Abu Mariam to smoke. Islamic State devotees were not supposed to smoke cigarettes, but Abu Mariam was addicted and wasn't giving up his habit for anyone, not even God.

Harith knew this was the time for him to act—he had to get to the phone before it was too late. He followed the grooms into the kitchen, pondering how to make his move.

Both men were gobbling up hard-boiled eggs and round pieces of flatbread left over from the previous night's dinner.

Bil Aafiyah, he told them, the Iraqi idiom for wishing someone a good meal. Take your time. I still have things to do before our journey.

He left them to their food and walked outside to the barn, waving his hand to Abu Mariam as he crossed the courtyard.

The commander looked Harith up and down.

Did my snoring keep you awake?

Harith forced a weak laugh.

No, Saidi, no. I slept soundly, ready to take on my mission. But now I'm in a hurry. I must get fuel. God forbid we fail because the car runs out of gas.

Inside the barn, Harith pulled the dust cover from the well-used white Toyota Corolla sedan, the car registered to Abu Suhaib. He opened the driver's door and sat down, repressing the urge to reach into the hole in the fabric near the floorboards for the phone that the Falcons' tech team had hidden there. Instead, he turned on the motor and deliberately kept both hands on the steering wheel as he drove out to the yard past Abu Mariam.

The commander pulled a pile of bills from his pocket, peeled off several for Harith, and leaned over to him in the car.

I'll be back in half an hour, Saidi.

Go with God, Abu Suhaib.

Harith whispered his thanks to God as he drove past the jihadis to the edge of the farmyard, then turned left down the dirt road that led toward the center of Tarmiya.

As the farmhouse slipped from view, a canopy of palm trees crowded out the sky. The road was empty as Harith trundled onward. He navigated potholes with his right hand on the steering wheel while he searched with his left for the rip in the car seat near the floor where the small black Nokia was supposed to be. When his fingers touched the plastic handset, he allowed himself a small smile of relief. Then he began to worry again.

Did he have enough battery power to make a call? He had no charger with him, and no time to charge the device anyway. He couldn't ask the gas station attendant to help him, because something out of the ordinary like that would be sure to work its way back through the local gossip chain to Abu Mariam. Unbidden, Abu Ali's mantra came rushing back to him. The small things, the details—these are the things that will keep you alive.

He needed to switch on the phone before he reached town, but first he needed to work it free from its hiding place. He did not dare stop the car in case anyone was watching him. Slowly, he worked the plastic handset through the cut in the fabric. It was more difficult than he had expected. Soon, the rows of date palms thinned and the empty car approached the intersection where he would turn toward town. Harith willed himself more patience and finally succeeded in pulling the phone free. He held his breath as he pushed the power button and let out a loud sigh when he heard the familiar jingle. The phone worked. He could connect with the Falcons to warn them of the pending attack.

Harith turned the sedan toward the gas station. Keeping his eyes on the road, he located the single number preprogrammed on the handset. After three rings, he cut the connection, just as he had been taught.

Then, he redialed and cut the phone call again after three rings. The third time he dialed, Munaf answered.

The family plans a delivery today, he told his brother. Two presents are expected to be en route before nine A.M.

The code was simple enough. The Falcons needed to prepare to intercept two bombers. And they needed to move quickly. It was almost six-thirty A.M. and Baghdad was an hour's drive away.

Munaf cut the line without answering and Harith quickly turned off the phone again just as he reached the gas station on the edge of Tarmiya's main road. There was no time to put it back into its hiding place without being seen. Instead, he kicked it under the seat when he opened the door to greet the attendant.

Salaam wa Alaikum ya Haji.

And peace be with you, my son.

I'm traveling today and need a full tank. Can you oblige me this request?

You are welcome, my son. My hands are at your service.

The elderly man stepped around the far side of the car to start pumping gas. The station was empty, other than Harith's car, and looked forlorn in the early morning light. The weather-beaten building where the attendant kept supplies was dark, with years of grime covering the window. As the attendant worked, Harith seized the opportunity to hide the phone again. He put his shoe on the running board of the car and pretended to tie his laces.

Praise to God. It should be a fine day for a drive, the attendant called over the top of the car to him. Where are you heading?

Harith looked up from his shoe. Baghdad, inshallah.

He quickly lowered his eyes. He didn't want to encourage a conversation. He pulled his shoelaces tighter and swiftly stuffed the Nokia back through the ripped seam in the carpet.

I can't imagine wearing those, the attendant said.

Harith looked up again, uncertain about what the old man meant.

Those shoes, the attendant said, pointing at Harith's feet. The laces cause trouble. With my back pain, it's painful to lean over and tie them the way you do.

Harith realized that the attendant had not been as distracted as he had thought. There was no way to know whether he had seen the phone, and if so, whether he would mention it to anyone. Harith would just have to act like everything was normal. He bent over one more time to pick up some lint and a cigarette butt from the car floor.

Thankfully my body is still strong, he said, staying out of the older man's view.

Harith threw the trash on the ground and handed the attendant the bills Abu Mariam had given him.

May God keep you safe and sound, the man said as he handed back Harith's change. Send my greetings to Abu Mariam.

Harith turned the engine back on and pulled away. For anyone else, the brief exchange would have been unremarkable. But living in the shadows, between the good guys and the bad, Harith could never be certain which side anyone else was on. Was the attendant just being friendly? Or was he threatening to expose Harith with his reminder that the commander was his friend?

Harith drove as fast as he could back to the farmhouse. When he arrived, he saw that the grooms had finished eating and were waiting for him in the courtyard along with Abu Mariam.

The commander didn't glance at Harith. Instead, he clapped both the Iraqi and Tunisian on their shoulders as a farewell. Go with God, he told them. May he guide your path.

Harith pulled away from the farmhouse for the second time that morning and drove along the same hushed, shaded road. His passengers sat silently in the back seat, their slender frames transfigured by bulky explosive belts that they wore underneath tightly buttoned winter coats. The early-morning air was cool enough to make their clothes appear appropriate. At least that was what Harith hoped, as he needed to get them past the first checkpoint without backup or aid.

After passing the gas station and turning onto the Baghdad highway, Harith rolled down his window. He needed some fresh air to lighten the mood. The passengers hadn't said a word since he started driving. He still didn't know their names. He didn't want to know.

He studied the men through the rearview mirror. The Tunisian sat directly behind him, his body rigid and tilted toward the window. His gaze was focused and distant, like he was thinking of someone he once loved. The hands of the second man were clasped, while his right foot tapped furiously. He seemed Iraqi through and through. From the previous night's conversations, it was clear he had understood the local idioms and contractions that befuddled other Arabs. His stubby fingers were scarred with cuts, like someone well acquainted with hard work.

The Iraqi met Harith's eye as the car trundled over

the broken, uneven pavement leading to the entrance ramp for the highway.

A thousand pardons, Harith said. The road will be smoother now.

The Iraqi returned his gaze. His dark brown eyes gave off a shine like a cornered animal.

How far is the journey? he asked Harith.

You've never been to Baghdad?

Never. My uncle lived there before the invaders arrived. But we never visited him. He fled when the infidels confiscated his home.

It's a pity you will never see the glories of Baghdad.

The man paused and looked out the window.

Inshallah my rewards in Heaven will be greater.

Harith felt a twinge inside his chest, something resembling pity. Then, he heard Abu Ali's chastening voice in his head: This man is not your friend. He would kill you if he knew who you really are.

Harith swallowed the lump in his throat and responded as a zealous believer would.

Depending on traffic we should reach our destination within two hours, he said. May God grant you the rewards that you deserve.

The Corolla moved steadily along the highway past rocky, shrub-choked fields and empty plastic bags dancing in the breeze. Since the start of war against the Islamic State, local farmers hadn't bothered to work this land that at any moment could become a new front line, or slip into enemy hands altogether.

Soon the cars ahead of Harith started to slow down at the approach of the first checkpoint. The right lane of the road quickly became clogged with eighteen-wheel trucks, the lifeblood of Iraq's retail economy, carrying all the goods that kept the capital fed, furnished, and entertained. Furniture and fruit from Turkey, electronics from China, rice from India.

The truck drivers sat with legs dangling out their doors, smoking lazily as the checkpoint soldiers attempted to conduct the traffic. They prioritized smaller vehicles driven by workers, while large trucks waited hours or even days to pass. Harith rolled up the car windows against the percolating black clouds of exhaust and merged into the fastest-moving line.

Harith could see four soldiers at the approaching tollbooth. They stood under the meager shade of an overhang decorated with the insignia of their military unit and with sun-faded posters of their members killed in the line of duty.

Stop fidgeting, Harith told the Iraqi, his voice harsh. You look nervous. We don't want to give them a reason to look inside the car.

The Iraqi looked abashed as the Tunisian glared at him.

This is not the time for second thoughts, the Tunisian said.

Don't question my fortitude, the Iraqi told him. You don't know what is in my heart.

Shut up, Harith told them both. No talking as we pass.

The queue moved swiftly, despite the crowded road, thought Harith, as he watched the soldiers wave the vehicles through without bothering with any inspection. The two guards were middle-aged and their uniforms bulged around the waist. Neither appeared to be fully awake or focused on the job at hand, which wasn't to move traffic but to watch for signs of danger.

Harith kept the Corolla moving and kept his eyes trained on the minivan in front of them, an intercity taxi whose driver was ferrying families from Anbar to their relatives in the capital. He hoped that by avoiding eye contact, he would avoid being stopped. The plan worked. The soldiers barely gave the Corolla a glance. In a blink of an eye, they were through.

Harith didn't notice anyone from the Falcons at the checkpoint. But as he pulled back onto the highway and picked up speed again, he hoped they had seen him.

Munaf had been half-asleep in Sadr City when he heard his brother's signal. He had worked late in the Interior Ministry's electronic surveillance center and only got home around two A.M.

His wife, as always, had risen for the dawn prayer, and she had moved his phone to their small sitting room to plug it into the charger there. He had told Nismah a dozen times. The phone was his most important possession. It could never be shut off and its battery could never run low.

When it started to ring at six-thirty A.M., Munaf woke immediately and scrambled to find it. The brothers' code was tailored to fit the layout of the al-Sudani home. Munaf's suite only had electrical outlets in the sitting room, not the bedroom, so it was always possible that he wouldn't reach his phone on the first attempt. Therefore, they agreed that Harith would try to contact him a second time if the first failed.

Munaf reached the phone the third time it rang. By the time Harith delivered his message, he was fully awake and already pulling on his trousers.

Nismah brought him a glass of sugary tea while Munaf began alerting the team. He ordered an armed sniper unit to deploy immediately. Then, he called Abu Ali.

A delivery is coming, he told the Falcons' commander. Possibly around nine A.M. Two grooms en route.

Call me back if they arrive in Baghdad, Abu Ali told him. God willing you will make sure they fail.

Munaf jumped into his black Mitsubishi sedan. He cursed as he wove in and out of the early rush-hour traffic. Even at this early hour, it seemed half of Sadr City's residents were already on the road. He pushed north to Baghdad's ring road, barking orders on the phone to muster the other four Falcons who, like him, were racing to prevent twin suicide bombings and to keep his brother and the city safe.

Munaf had already helped foil seven separate bomb attacks and in the process he had honed a procedure for dealing with these threats. While he sped along the highway, slaloming around the heavy trucks and packed minibuses, Munaf pulled into the Falcons' headquarters in record time to join the rest of the four-man team. Time was precious. They had less than an hour to get into position to intercept Harith's vehicle. The men headed north out of town. Their destination was the army checkpoint closest to Tarmiya, part of the multitier defensive belt surrounding the capital. The system created traffic nightmares but in theory they were supposed to provide multiple ways for the security forces to stop terrorists trying to penetrate

Iraq's largest city. The weakness of the system was the inconsistency among the different branches of Iraqi security forces that controlled the checkpoints. Some units allowed every car through without as much as a glance. Others demanded bribe money. The Islamic State commanders who recruited Abu Suhaib into their courier network had done their homework, taking advantage of the system's flaws. With his Baghdad identification card and license plates, he was unlikely to be stopped. If soldiers demanded a bribe, he would have money to hand over. Even in the far rarer instance when passengers were asked to step out of the car to be frisked, the militants had a plan. Under those circumstances, the militants would detonate their explosives then and there.

Munaf and his team were the only bulwark against such a tragic outcome—either that the bombers would reach Baghdad safely and execute their original plan, or detonate themselves before that, killing Iraqi forces and Harith.

Munaf received a progress report around 8:45 A.M., as he and his team reached their destination. Harith's white Toyota Corolla, he was told, had passed through the outer checkpoint with two passengers in the back seat and was heading down the Abu Ghraib highway toward Munaf's location.

Munaf quickly ordered one of the Falcons' sharpshooters hiding on a shrub-covered rise overlooking the road to get ready. The sniper was as familiar with the terrain as the local farmers were. He knew there would be only a few minutes before Harith's white Corolla came into view.

The younger al-Sudani officer, meanwhile, demanded to speak with the officer in charge of the checkpoint. He explained that the Falcons had a special operation underway to neutralize a bomb attack, but he didn't mention Harith or the undercover mission. To avoid bloodshed, he urged his fellow officer to keep his men away and let his unit handle the situation. The last thing Munaf wanted was a panicked suicide bomber killing fellow soldiers, or the checkpoint militia shooting his brother.

As he was talking, the sniper broke in on his radio. Corolla approaching, he told Munaf.

Wait for the signal and do what you must, Munaf replied.

Harith kept the Corolla in fourth gear along the highway. The battered vehicle had a strong engine, but it wouldn't win any beauty contests, nor any races, not with its current transmission. Still, the car was perfect for the job. He and his passengers' deadly payload blended into any situation, invisible to anyone who wasn't expecting them.

A few hundred yards before the bend that led them to the checkpoint where Munaf waited, Harith started issuing directives.

Listen closely, he told the bombers. The next checkpoint is around this curve. And it won't be as easy as the first one we passed. This security unit is serious. They often have bomb-sniffing dogs and they stop cars for a complete search.

None of what he was saying was particularly true. But it wasn't particularly false, either. The diligence of the checkpoint commander dictated how soldiers treated traffic. But the bombers didn't know that.

Now you must do exactly what I say, Harith told them. In a few minutes, I am going to pull the car over to the side of the road and you will get out. I am going to open the hood of the car like I'm having engine trouble. And you will leave the car and start walking to the checkpoint. People on foot, either in the fields or along the road, are never stopped. And that is exactly what you will do. You'll walk and, in a few minutes, I'll restart the car and drive through the checkpoint after you. That way if they stop me and search the car, they won't find anything. I'll pick you up again on the other side.

The grooms looked at Harith. No one had mentioned any of this before. But there was no reason for them to question what he was telling them. And there was no one to confirm his instructions, even if they had doubts.

Harith slowed the Corolla and pulled over onto the shoulder along the western side of the road. He opened his door and got out, walking slowly to the front of the car.

Go now, he told the bombers. Start walking.

The Tunisian and Iraqi did as they were told without complaint.

They passed Harith as he reached for the latch to open the hood of the car.

Go with God, he told them as he lifted the hood and propped it up into place.

Harith leaned over the Corolla, appearing, for all intents and purposes, like a driver with engine trouble. Like the night before, his nerves were jangling, and he willed himself to be still. This was the sign that Munaf's team was looking for, telling the Falcons to swoop in fast.

Time seemed to stop and Harith later told Munaf that he became dizzy from a sudden rush of blood to his head. His heart was pounding so loudly that he didn't remember the series of cracks that split the air behind him and the thuds of two bodies falling on the road.

The first thing he remembered hearing was a whooping from a camouflaged figure rising from the scrub bushes blanketing the east side of the road. Mission accomplished, the sniper yelled.

Munaf sprinted toward his brother, yelling at him to drive away from the scene. The rest of the Falcons team, the men trained to dismantle the suicide bomb belts, were trotting along behind him, keen to ensure that the explosive devices were disabled.

Harith felt his legs wobble. He turned away from the car and looked at the two bodies lying about fifty feet away. Half of the Iraqi man's head had been blown off by the sniper's kill shot. The Tunisian was facedown, blood pooling around his torso. He didn't appear to be breathing.

Harith spent the next six hours with his brother back in Baghdad, debriefing him on everything he had seen and what he knew of the Islamic State commander in Mosul, including the phone number that he used and his encrypted username on Telegram. He didn't tell Munaf about his nightmare.

He then slept deeply for three hours, his body drained. Munaf shook him awake, fed him, and ordered him back to Tarmiya.

We have already released the news to Iraqi television that fresh attacks were carried out, he told his older brother. The Islamic State will believe your mission was successful. If you don't return now, they will wonder what has happened to you.

Sure enough, two hours later when Abu Suhaib pulled back down the rutted dirt road to the farmhouse, the entire cell gathered to meet him, chanting a prayer of victory.

Allah bless your hands, Abu Mariam told him as he climbed out of the car. Thirty unbelievers died today.

Harith covered the car, washed up, and joined the jihadis in their prayer of thanksgiving.

Chapter 18
Caught in a Trap

As far as the Islamic State was concerned, by the summer of 2016 Harith had become one of their most successful Iraqi recruits. Since joining the cell in Tarmiya, the Islamic State military commander in Mosul, Abu Qaswarah, had assigned him sixteen missions aimed at the Iraqi capital. Half the missions involved fertilizer bombs packed into vehicles that were designed to detonate around Baghdad: shopping centers; police stations; and the large public park near Sadr City. The other half were human suicide bombers, people like the Tunisian and the Iraqi, whose aim was to blow themselves up in as crowded a location as possible.

As far as Abu Qaswarah and Abu Mariam knew, each time Abu Suhaib left the farmhouse on a mission, he had a 100 percent success rate. Not only had he never been stopped at a checkpoint, but also the bombers he ferried into Baghdad had never been arrested. The Islamic State leaders believed that their soldiers died as martyrs, killing dozens of unbelievers in the process. The truck bombs manufactured in the group's production factory in Al Qaim never failed—they destroyed lives and property and sowed terror in the capital.

The work Abu Suhaib had ostensibly achieved in Baghdad was a relative bright spot for the Islamic State. Elsewhere across their self-defined caliphate that summer, U.S.-backed forces had overrun their positions in the Syrian city of Kobani after weeks of intensive fighting, while Iraqi armed forces had regained control of Fallujah, located some forty-five miles north of Baghdad.

The militants were also reeling from a string of successful assassinations of several major Islamic State figures, including their so-called oil minister in 2015 by U.S. special forces. In March 2016, a coalition airstrike killed the Islamic State's minister of war, the notorious Chechen fighter known as Abu Omar al-Shishani. Then, another leading battlefield commander, a veteran Tunisian jihadi, was killed in another airstrike. In July 2016, the terror group still controlled a territory the size of the United Kingdom and had approximately four million people under its control, but its veneer of impregnability had cracked. The Islamic State no longer seemed unstoppable.

Inside the Baghdad Operations Command—composed of Iraqi army, intelligence, and frontline police commanders as well as military and intelligence officers from the international coalition—there was cautious optimism that the tide of war was shifting in their favor. But the price of those victories had been high. Almost 30 percent of Iraqi men of fighting age had been mobilized, and Iraqi military units were suffering casualties at loss rates that would be unacceptable in any other country. Funerals were so frequent that the country's major cemeteries were undergoing rapid expansions to make room for the bodies.

Abu Ali kept telling his colleagues that there was a ray of hope. Baghdad had been almost completely immune from terror attacks for several months. Thanks to the Falcons, Islamic State sleeper cells had been rolled up in the capital. Where once the city had suffered daily terror attacks, there had been only three verified suicide bombings in six months. The Falcons had press releases issued for each of the bombings that Harith had been ordered to help execute as a method to buttress his cover story, but in reality the city—where more than one-fifth of the country's population lived—was safer than it had been in years.

No one at Baghdad Operations Command knew of the Falcons' undercover officer or the classified missions the unit was conducting to keep the capital safe. But everyone could see that, without a doubt, morale among Baghdad residents had changed. Two years earlier, Iraqis feared that Baghdad would be sacked in a repeat of the thirteenth-century pillage of the city by the Mongols, yet by the end of 2016 residents didn't fear for their lives each time they stepped out of their front door. True, the nation was still under occupation, but families were sending children to school, and some businesses were expanding.

Abu Ali al-Basri knew that secrecy was crucial to Harith's ongoing success, so he never created a paper file about his mole. Instead, he made sure the prime minister understood that the Falcons had a potent weapon in the field as part of the effort to defeat the Islamic State. In select briefings, the spymaster would refer to Agent 31, the code name he used for his undercover officer when he shared crucial information about the militants gleaned from Harith's almost daily conversations with Abu Qaswarah. The Americans had the capital of the caliphate under electronic surveillance, but the Iraqi government never knew how much information they shared, and what they withheld. For seven months, Abu Ali handed the Iraqi prime minister high-grade intelligence from top Islamic State masterminds, something that no Iraqi leader had previously had at his disposal.

Each time Munaf debriefed his brother, the spymaster reminded him to convey his gratitude for Harith's hard work. The prime minister and generals may not know your brother's name, Abu Ali told Munaf. But they know the sacrifice he is making. Tell him that we all salute him.

These meetings with Munaf were Harith's only contact with the world outside the Islamic State, a fleeting respite from the rigors and stress of his work. That summer, after yet another foiled suicide bomb attack, Munaf sat with his brother in a safe house near the Abu Ghraib prison running through what had become a familiar template of questions. What had Harith talked about with Mosul? Who had visited the farmhouse? How was Harith feeling? Munaf could see that his brother's eyes were red from exhaustion and that he had lost weight. When he relayed Abu Ali's praise, his normally reticent brother started to cry. Looking back, Munaf realized that surprising display of emotion was a tip-off that something serious was bothering Harith. But at the time, he didn't think to ask what it was.

In Mosul, the Islamic State's setbacks on the battlefield in mid-2016 weighed heavily on the group. Among the messianic wing of the movement, a sinister point of view was taking shape to explain the loss of momentum and the elimination of some of their most effective men. Surviving commanders became paranoid. In their fevered belief in holy war, there could only be one reason for these setbacks: treachery.

One of the common misperceptions in the West about the Islamic State concerned its demographic makeup. The number of foreign Muslims from Western countries who had fallen for the group's slickly produced propaganda campaigns were vastly outnumbered by the number of local Iraqis and Syrians, as well as other Arabs, who made up the backbone of the group's bureaucracy and its fighting force. The group's leaders were predominantly Iraqi, men weaned on the mind-bending brutality of Saddam Hussein's regime and his torture squads.

It was no wonder, then, that the men running the Islamic State mimicked the dictator's methods. They established multiple, overlapping layers of security police, legions of informants who spied on their neighbors and all public interactions to ensure the rigid rules of the caliphate were being obeyed. Whether the length of a man's beard or the prohibition on smoking, the smallest of infractions demanded severe punishment. Imagine, then, what these enforcers would do to a suspected traitor. When the battlefield setbacks cascaded, the caliphate's administrators, whether a town magistrate or provincial governor, were on the lookout for spies, those wreckers who were sabotaging the vision of their religious state. No one was above suspicion, not the brilliant scientists who had flocked to the Islamic State from countries such as France or Turkey, nor the frontline operatives who were smuggling soldiers of God into Baghdad. No matter how dedicated or successful a devotee had shown himself, the Islamic State was predisposed toward suspicion.

In Tarmiya, Abu Mariam followed Mosul's dictates to the letter. His actions were in keeping with the type of person that Harith had described in his debriefings. Ascetic and slender with a ramrod straight back, and deeply creased skin, the farmer-turned-jihadi had a severity of manner that made it clear that orders were to be obeyed without question, a demeanor not dissimilar to what Harith experienced growing up in Sadr City.

Harith never knew precisely why Abu Mariam supported the Islamic State. He was a religious man, of that there was no doubt. But, like most Iraqis, he had not grown up with the restrictive, puritanical interpretation of Islam that the terror group espoused. He did, however, have a sharp contempt for the Shiite politicians who had taken over the country after 2003, and like other rural Sunnis, he had found it impossible to adjust to the new political realities after Saddam's ouster. From his point of view, there was honor to be won by resisting the new government. Better than being reduced to a life as a second-class citizen. If that choice meant killing fellow Iraqis, including women and children, so be it. His worldview was Manichean—it was him and his kind versus everyone else, and Iraq's Shiites and the Americans had proved they were on the other side.

Victory in the battle for survival required discipline, rules, and punishment, and soldiers—even those whom he believed loyal to the cause—needed reminding of the price for betrayal and failure. So during that summer of 2016, Abu Mariam filled the hot, dreary afternoons at the farmhouse with mandatory viewings of Islamic State propaganda videos. With the ceiling fan churning enough air to cool the living room, his men sat on the floor of the large room, legs crossed or reclining on cushions, around the laptop monitor to watch the raw and violent videos produced as part of an increasingly frenzied hunt across the caliphate to root out traitors within its ranks.

One video showed men denounced as spies tied to a cross in the Islamic State's Syrian capital of Raqqa and then shot dead in front of assembled onlookers. Another video included taped confessions of men who admitted to receiving payment from the Islamic State's enemies in exchange for providing targeting information to coalition forces. These men were then chained inside a car, which Islamic State soldiers then detonated. Another film discussed a female member of the caliphate castigated as an agent for Moscow's intelligence agency after the deaths of several militants whom she had known.

Abu Mariam kept the clips running nonstop until the farmhouse generator shut down for the night. With no other entertainment available, it was rare that the men would turn away from the grotesque films, even after repeated showings. After all, they depicted a version of reality that the men in Harith's cell wanted desperately to believe. The group to whom they had pledged allegiance was on a righteous mission and defeat would not be tolerated.

Harith worked hard to remain impassive as he watched the executions of the traitors over and over again. He wondered what exactly led each victim to confess the alleged crimes. He presumed the men had been tortured, but as he scanned their bodies for signs of mistreatment, he never saw any obvious marks of beating or abuse. The Islamic State cameramen were skilled at showing only select parts of a prisoner's body. A face free of cuts or bruises would be shown on camera, for example. But not the torso, arms, or legs. No matter how skilled the production quality, the propaganda teams could not edit out the fear in the eyes of the prisoners, or the tight grimace on their faces as they recited their scripted confessions.

That summer, Islamic State political commissars spent a night at the Tarmiya farmhouse after a mission farther west, to Anbar Province, to root out suspected spies. A couple months earlier, a veteran Iraqi militant, Shaker Wuhayeb, had been killed in a Coalition airstrike, fueling the panic about possible double agents working against the terror group in western Iraq. The commissars forced the Tarmiya cell members to watch a new video showing the execution of two dozen members of Wuhayeb's tribe on suspicion of feeding GPS coordinates to the Iraqi military. It was an open secret that the Iraqi military was corrupt. Islamic State fighters were allowed to pass through the front lines completely unmolested for the right amount of cash. But it was a source of fury among Islamic State leaders that men from within the group's own fighting ranks would take a bribe or betray them.

The commissar ended his lecture with a chilling slogan. The path to victory needs a gun with ten bullets, nine for the traitors and one for the enemy, he said as he drank a cup of unsweetened tea. May they all rot in hell.

At the start of July, Harith received a slew of good news. He had passed what in essence was his first job performance review with the Islamic State. Abu Mariam praised his dedication and remarked on his successful missions. Soon after, during his daily phone call with Abu Qaswarah in Mosul, the commander told him that he was being promoted and given more responsibility.

Instead of his $300-per-month salary from the terror group, he would start receiving $600. And in addition to ferrying suicide bombers into the capital, Harith would now help select possible bomb targets as well.

Harith couldn't believe his good fortune—not only would he be in a position to stop bombers from attacking the capital, he could direct the list of targets, information that he could give the Falcons well in advance so that they could concentrate surveillance on specific sites.

What he also realized, as he drove away in his battered Toyota from the farmhouse toward Baghdad on his first such reconnaissance mission, was that these new duties would give him ample time to meet Munaf. He could fully brief his true band of brothers, the Falcons, instead of the rushed messages that he sent on Telegram on the rare occasion he could access his secret phone. He might even be able to smoke a water pipe or see his family. It was the holy month of Ramadan, a time when Muslims fasted by day and then feasted at night with their extended family. Harith was curious about how his children were doing, how much his son had grown in the six months they had been apart.

Harith had no deadline for returning to the farmhouse. Abu Mariam knew he had been given a vital mission and, as far as Harith knew, his commander trusted him, despite the hysteria over spies. After all, he had received nothing but praise for his work.

By the time Harith passed the two checkpoints on the outskirts of Baghdad, he was overwhelmed by the idea of going home. He could break the Ramadan fast, eat his mother's food, see his younger brothers and his children. Surely, they must miss him as much as he missed them. And maybe it was time they understood just how important he had become.

Once Harith turned onto the Baghdad airport highway, the six-lane road that headed toward the center of the city, he pulled his secret phone out of its hiding place and called Munaf.

Brother, I'm in Baghdad. But don't worry. There is no planned attack. I have a new mission. One that will please you, Harith said.

Munaf was surprised, but pleasantly so. His curiosity was piqued when his brother asked him to meet at their father's house at sundown.

You're going to Sadr City? Munaf asked, astonished.

Brother, don't worry. I'll explain when I see you.

He then called their youngest brother, Munthir, telling him to inform Um Harith that he would be home to break the Ramadan fast.

I have a short break from my mission, he said. So tell her not to make a fuss. I can't stay long.

Two hours later, Harith was sitting in the al-Sudanis' living room at his usual place around the sofra. His mother served steaming platters of fresh butter beans and rice, boiled chicken, and cold glasses of laban, the sour yogurt drink. All eight brothers had been alerted by their mother as soon as she learned that her oldest son was arriving. As the others returned home, they came to greet Harith like an honored guest, kissing his cheek in the elaborate fashion Iraqis reserve for close friends. All they knew was what Munaf had told them: Harith was on a top-secret assignment vital to national security. Harith's own children were a bit reserved, each wondering where their father had been all this time but knowing it was not their place to ask.

The family was still in the midst of exchanging their pleasantries when a phone rang, momentarily interrupting the happy commotion in the room.

The al-Sudanis couldn't believe their ears. The ringtone was the tune of a common Islamic State chant, the same religious dirge that the group often used in its propaganda videos. No Shiite in Iraq would have such a tune saved on their phone. Was this some sort of joke that Harith had with his friends at work?

Without a word, Harith fled the room, locking himself in the bathroom before he pressed the button to answer his phone. The ringtone programmed on the handset that he used as Abu Suhaib was for his most important contact. Abu Qaswarah, his commander from Mosul, was on the line.

God be with you my sheikh, he said, remembering to adopt the choppy accent of Abu Suhaib.

And with you, Abu Suhaib. I'm calling for an update about Baghdad, his commander replied. Where are you, my son?

Harith felt an icy blast of dread come over him. Had he been followed by an unknown Islamic State unit during his reconnaissance mission around Baghdad? Were they watching him even now?

He answered with the first thing that came to mind.

My sheikh, I am in Baghdad as you ordered me. I'm in the neighborhoods of the rafidah, he told his commander, using the word that the Islamic State used for Shiites, the rejectors, the ones who turned their backs on what the group regarded as the true interpretation of Islam.

I'm looking for the best targets for our soldiers, as you ordered, Harith continued. Do you have any new instructions for me?

Abu Qaswarah paused before replying.

No, my son. As long as you are fulfilling your orders, I have no other wish for you, he said. Give Abu Mariam a full report when you return this evening.

After the commander hung up, Harith realized that his back was wet with sweat and his heart was beating like a wedding drum. He took several deep breaths before going back to the crowded family room.

In the few minutes he had been away from the room, Munaf had arrived home. At first glance, it seemed to

Munaf that his older brother had aged ten years. Before Harith could say a word, Munaf took charge.

Mother, he said. One thousand pardons, but I have been ordered to fetch Harith and return him to headquarters. There is an emergency and he is needed urgently by our commander.

The words worked as Munaf knew they would. As Munaf marched Harith out to the courtyard and back to their cars, their parents suspected something strange was going on. It had been months since they had seen Harith, and then out of the blue he popped home only to be rushed away again. But they didn't ask questions. After all, there was a war on.

Ya khara, Munaf said as the two drove away. You shit. What has come over you?

Do you think you are on a vacation? Who gave you permission to come home?

Harith looked haggard, like he had the flu. Munaf could see that something terrible had happened. He just didn't know what.

Munaf, I think I'm in trouble, Harith finally said. Abu Qaswarah may suspect that I'm a traitor.

Munaf pulled the car over by a roadside tea vendor as Harith began describing his insomnia and anxiety. He told his brother about his job evaluation and promotion and the growing paranoia among his Islamic

State comrades. He described his nightmare and the unbearable sensation of being pulled underwater and drowning.

Munaf was furious.

Why are you tempting the devil? Munaf asked him. Why, when they show you their trust, do you run the risk of disobeying them straightaway? Don't you see how reckless you have been?

You have to pull out. You aren't thinking clearly.

Harith fell silent in the car as Munaf drove them westward, back toward the highway to Tarmiya. Earlier in the morning, his plan hadn't seemed risky. Was he losing his edge after six months undercover?

Harith refused his brother's demand. He told Munaf he would return to the farmhouse and pleaded with him not to tell anyone about his indiscretion or his nightmare.

He told Munaf that the phone call from Abu Qaswarah had shocked him straight.

Don't worry, he told Munaf. I'm under control. By the time I reach Tarmiya tonight, all they will see is Abu Suhaib. There will be no more mistakes. I promise.

Munaf saw his brother's determination and finally gave in. God forbid Abu Ali finds out about this misadventure, he thought. Cautious and phlegmatic as he was, the Falcons' chief might insist on pulling Harith from the field.

Harith told his younger brother he wasn't going to allow that to happen. He wasn't going to throw away his shot at glory.

Two days later, a van packed with explosives drove down the crowded streets of Karada, an upscale neighborhood along the Tigris River close to the Baghdad University campus and several government ministries. Some of Baghdad's most expensive shops lined the district's historic streets where literary legends and intellectuals sat in sidewalk cafes to drink tea and discuss the state of the world.

It was the end of the month of Ramadan and families thronged the stores, searching for holiday outfits to wear during the Eid celebration that was set to begin. Young men flocked to the new sports bars and shisha cafes that had cropped up thanks to the fresh sense of stability around the capital. The European championship tournament was on and the soccer-mad fans were watching the quarter-final matches.

The driver of the nondescript van drove by several outdoor cafes and detonated his payload outside a popular three-story shopping mall, killing a small group of bystanders instantly. The blast set alight anything flammable. In an instant, the mall became a roaring inferno, trapping hundreds of people inside.

Baghdad was in shock. By the next morning the death toll stood at more than three hundred people, making the Karada bombing the single most deadly attack in Iraq in more than a decade.

Prime Minister Haider al-Abadi convened an urgent meeting of his security chiefs while emergency workers were still trying to put out the flames and pull bodies from the charred wreckage of the mall. He couldn't contain his anger at the tragedy. Baghdad had been on high alert, a precaution against the Islamic State's known penchant for attacks during the Islamic holiday season.

How the fuck had terrorists penetrated the heart of the capital, less than half a mile from his own family home, the prime minister shouted at the generals and ministers sitting around his conference table.

No one had answers, not even Abu Ali al-Basri. In that moment of white-hot anger, it was no consolation that he and other members of the Iraqi security forces had successfully foiled dozens of other planned attacks. In their line of business, there was no room for error because even one failure was one too many.

As the prime minister raged, Munaf was on the other side of the river at the Falcons' headquarters, logged into his encrypted platforms trying to figure out what had gone wrong. He hadn't missed any communications from Harith, or any chatter from any of the Falcons' other agents about a large-scale attack on Baghdad. He had nothing to report to Abu Ali when his boss called for information, any details, that he could give the prime minister.

Televisions across Iraq broadcast wall-to-wall coverage of the blast's aftermath. Among the victims of the fire was Adel al-Jaf, a young dancer also known as Adel Euro who had been set to start a fellowship in New York. Zulfikar Oraibi, the son of a former Iraqi football star who had played in the 1986 World Cup, was also killed. Many of the victims were children who had been playing in the mall's arcade.

Two days later, the prime minister fired the interior minister and the head of the national intelligence agency. Abu Ali was promoted to the head of counterintelligence and head of national security.

One of the spymaster's first acts in his new role was to deliver a report compiled with the help of Iraq's domestic intelligence agency. They concluded that the driver had traveled to the capital from Diyala Province, near the front line between the Islamic State–held territory and Iraqi federal land. He drove down the Salahuddin highway to reach Baghdad, nowhere near Tarmiya or Harith's cell.

What Abu Ali omitted from the report was the fact that his undercover officer had been frozen out of this operation. Which was no comfort to the spy chief. Since Harith was inserted into the enemy ranks, he had been informed of all major operations targeting the capital. The rigid bureaucracy of the Islamic State dictated that members acted within strict boundaries of their own territory, and the Tarmiya cell was in charge of Baghdad.

So what exactly was the enemy doing now, Abu Ali wondered. It never occurred to him that Harith—Abu Suhaib—was being watched, or that his loyalty was under question.

Chapter 19
Coming Home Again

The idea was that she would glide invisibly, like a ghost, from the airplane and through passport control, retrieve her drab suitcase from the carousel, and go home. It was late September 2016 and Abrar had her plan all worked out.

For the past several months she had been living in Turkey, keeping a low profile, like so many Arab-speaking refugees from the war zones of the Middle East. One part of the cover story she had told her parents actually came true. A Syrian director of a pharmaceutical warehouse had hired her off the books, which gave her some money to supplement the savings she had withdrawn from the bank before leaving Iraq. Outside of work, Abrar mainly kept to herself, rarely straying from her room at a shabby hostel reserved for women workers. She spent hours online in her familiar chat rooms with Islamic State supporters. Despite her disappointing trip to the caliphate, Abrar remained a believer in its goal of purifying the world of unbelievers. Swept up in the fervor of these online friends, she began thinking about staging an attack on her own, something she could do in the name of the Islamic State and its leader, Abu Bakr al-Baghdadi.

With the help of pharmaceutical supplies she pilfered from work, Abrar secretly rebuilt a stockpile of ricin, the toxin she wanted to use for her lone-wolf operation. By the start of fall, Abrar had a target in mind and a large enough supply to act.

Abrar messaged her family in Baghdad, telling them she needed to come home. Life in Turkey was too expensive and difficult for an Iraqi student, she said. Her parents were overjoyed to hear from her. Professor al-Kubaisi sent her money for a plane ticket. They expected her to arrive September 21.

The flight from Turkey to Iraq posed a serious challenge, however. Somehow, Abrar would have to pass layers of airport security without anyone raising suspicions about her lethal cargo, which she had concealed in empty cans of powdered infant formula. If she made it past the Turkish border guards at the airport, she would then have to lay low through the flight so that none of her fellow Iraqi passengers on the plane would remember her or alert the mukhabarat when they landed in Baghdad.

Her aim was to find a family to sit next to on the plane so she could blend in. Even in her high-necked shirt, black chador, and tight headscarf that concealed her body, Abrar knew that every man on the airplane would be eyeing her. Most would be discreet, but they would wonder who she was and why she was traveling alone. Religious women in Iraq rarely flew alone, unless they were wealthy, a status that guarded them from judgment and reproach. The last thing Abrar wanted was to find herself amid a public scene. Sitting next to a group of women would be a mistake, because Iraqi women liked to chat. They would prod and poke until they knew her whole life story even before the plane took off. No, it was better to sit with a family, where the mother would be too preoccupied with noisy children to bother to speak to her and too tired to take in what she looked like.

And, God be praised, that is exactly what happened. Abrar changed seats to be in a row alongside a young mother and toddler whose husband and two other children sat one row ahead. She gathered from the couple's conversation over the seat that they had been to Turkey on vacation. The couple spent the entire flight complaining about their hotel and slapping at their squirming children to sit still. Abrar, meanwhile, sat rigid and vigilant, ever aware of the black shiny leather satchel she had beneath her feet. Its contents were another reason she had secured a seat near a family. If anyone asked why she was carrying cans of infant formula, she would say they were for the children.

Once the passengers disembarked from the plane, Abrar knew it would be easy to hide in plain sight. Iraqis have no discipline about standing in line, and at passport control the security officers never tried to herd the crowds into order. As a result, women generally congregated in one line, nearest the senior official on duty, to escape the crush of bodies. While segregation by gender is not an Iraqi custom, courtesy and gallantry toward women is. The duty officer would generally open a special line for mothers traveling with children, sensing that they were tired and near their wits' end. Abrar hovered near the mother and children whom she had sat next to during the flight.

When it was her turn to hand over her documents to the immigration officer, she was just another mousy woman in a noisy, crowded hall. Nobody to notice, in case someone was watching.

Salaam wa Alaikum, she greeted the officer as she reached up almost above her head to pass him her passport.

He flipped through the document, the pages empty except for her Turkish entry stamps.

What took you to Turkey, my sister? he asked Abrar, looking at her for the first time.

I've been studying there, she replied, looking down, like a good Muslim woman should. Inshallah I will finish a degree as a cancer researcher.

God is great. He made you very small but with a clever brain, the officer replied. You must make your family proud.

They will be pleased when they see how I will help humanity, Abrar replied, jutting her chin out proudly.

The officer thumped his seal in an ink pad and stamped her entry back to Iraq. September 21, 2016. Welcome home, my sister. Go in peace.

Abrar walked calmly past the passport control lane and to the small X-ray machine used to scan hand luggage. Nothing in her bag should set off the machine, she thought. But if they decided to open her canisters of infant formula, the guards would be in for a surprise.

The queue in front of Abrar was hectic and jumbled with children throwing toys and backpacks on a small chest-high conveyor belt. Older men tried to maneuver through the crowds and hurry past. Abrar's lumpy bag disappeared through the X-ray screen and she tried to keep her balance as she was pushed along.

The chaos of the crowd annoyed her. A woman would never have to endure this in the caliphate, she thought.

Thankfully, no one in a uniform gave Abrar a second glance. She scooped up her satchel and, after a few minutes' wait, retrieved her larger suitcase from the carousel and walked past customs and through the double set of sliding glass doors without being stopped. She went past the grimy cafe that served piping-hot tea and the cell phone kiosk out to the street. Standing on the curb, waiting for a taxi with Baghdad's familiar dust-choked air swirling around her, Abrar said a quiet prayer of thanksgiving.

Everything was going according to plan.

Or so she thought at the time. In fact, the Falcons knew the date that Abrar was traveling back to Iraq and her flight information. Well before she landed in Baghdad, they had an operation ready to intercept her.

The Americans were aware of her too. After Islamic State fighters attacked Iraqi Kurdish forces in August 2015 with mortars containing mustard gas, an outlawed chemical weapon, the U.S. military had assembled a special unit to hunt and destroy the depots of toxins and chemical weapons that they knew the Islamic State had been building in Iraq and Syria. The customized mortar used by the militants caused painful blisters, burning, and nausea among its victims. It also instilled terror among the Iraqi Kurds who had survived the gruesome chemical weapons attacks ordered by Saddam Hussein against them in the 1980s. That war crime caused tens of thousands of deaths and led to stricter international bans of chemical and biological weapons. Iraq's Kurds were adamant that they would not allow that to happen again. In Washington, the Obama administration made the elimination of the Islamic State's chemical weapons division a top priority.

By the middle of 2016, the Americans believed they had eliminated all of the Islamic State's major chemical weapons manufacturing facilities in Iraq, in a series of airstrikes using munitions designed to destroy toxins and prevent them from spreading in the air. Their intelligence showed that the terror group had focused much of its chemical weapons production on mustard gas, because it was relatively easy to weaponize and could sicken or kill large numbers of people.

Abrar may have been rejected by the Islamic State's chemical weapons division, but she must have made an impression on some of the group's scientists and leaders. She was included on a terrorist watchlist. Abrar's relationship with Abu Nabil—the Islamic State leader who had been killed in an American airstrike in Libya 2015—also made her a person of interest to the Americans. When it became known she was returning to Baghdad, the Americans called Abu Ali for help. The spymaster said he would look into it.

The Americans had given Abu Ali a succinct briefing. They didn't tell him the extent of their knowledge about the Islamic State's scientific research program, or details of their previous operations targeting the group's scientists. What they told him was more circumscribed—a petite woman from a well-regarded Baghdadi family may pose a threat to the capital.

At first, the spymaster was cautious. The Falcons had no authority to detain Iraqis without evidence, and the Americans could provide nothing that he could take to a judge to request an arrest warrant. Besides, in 2015, Prime Minister al-Abadi had given strict orders to Iraqi commanders to keep secret any information about the Islamic State's unconventional weapons capabilities. What the Americans were doing was asking him to step into a political minefield, and he wasn't willing to do it. At least not without gathering more evidence first. So he ordered the Falcons to do what they did best. Set a trap.

That's how it transpired that Abu Ali's men had their eyes fixed on Abrar as she emerged from the airport and got into the taxi which drove her home.

For the next eight days, Abrar became the Falcons' most urgent assignment. Unlike Harith's mission, which was known to only a handful of people, Abu Ali briefed most of his men about the threat she posed and assigned two full units to watch her house and her movements around town. Additionally, he had men monitoring her phone calls and internet activity. He didn't know what she was up to at the start of the week, but by the end of it, he understood why the Americans had considered her a high-value target.

As Abrar settled in at home, her parents seemed more cloying then she remembered. During her months on the road, waiting in empty rooms to be called for consultations with Islamic State commanders or driven to another part of the caliphate, the twenty-nine-year-old woman had grown used to being alone. But she had missed certain things about home. The luxury of a hot shower and the soap imported from Dubai that she liked to use. Her mother's recipe for Margat Bamiya, her favorite lamb okra stew.

But Abrar hadn't missed the incessant talking, the endless debates about what a certain politician said on television, and the simmering but impotent anger expressed by her father and uncles. They didn't like what was going on in Iraq. But she seemed to be the only one willing to do anything about it.

For the first two days she was home, her parents pampered her like they did when she was a child. Her mother made her favorite meals and they sat for hours telling her the latest news about her cousins and aunts. Of course they wanted to know everything about Turkey. She told them what she thought they wanted to hear. She had enrolled in classes and found the studies easy. She had friendly and pious roommates who kept their room clean and never acted improperly. She found the cost of living too high and grew tired of juggling her academic work and her lab work.

Will you try to return? her father asked.

Maybe I will see if I can enroll in my old program again, Abrar told him. Maybe it's best if I stay in Baghdad.

After her parents went to bed, Abrar went online. Bint al-Iraq, Daughter of Iraq, rejoined her online chat groups where she had found support and encouragement before she journeyed to the caliphate. But the tone had changed. It was coarse and combative. What's stopping you? the Islamic State fanboys goaded her.

The caliph has commanded us to act in whatever way possible to attack the unbelievers.

I need more people, Abrar wrote. I need help to deliver my present.

The most dismaying aspect about life in the caliphate had been how boring it was. For weeks, she had had nothing to do. None of the things Abrar was used to doing in Baghdad were allowed. She couldn't walk outside without a male guardian, she couldn't work outside the house, she was forbidden from going online. And when she finally reached Mosul and asked the chemical weapons department for the raw ingredients she needed to manufacture and stock her ricin, she was told they weren't available.

In her hours of boredom came the first glimmers of the plan to attack in Baghdad. She fantasized about taking her ricin back to the ministry where she used to work and injecting it into her former colleagues' teapot. It would only take a small amount, the size of five or six grains of rice, to kill those ladies who for years had flaunted their polluted religious thinking at her. The thought animated her, but it wasn't ideal. Her former coworkers would not start showing symptoms until they were home at night. If her poison worked, each would be dead within a day. But they would never know what killed them. They wouldn't know it was her.

Abrar put that idea aside—she hadn't joined the caliphate to exact personal revenge. She had joined to purify Iraq of its sinful ways. Baghdad was full of people whose whole existence was blasphemy. She had a chance to rid her city of that filth and she thought she knew precisely what to do.

She implored her online friends once again.

I need two, maybe three, volunteers. Holy warriors who can help with my operation. People ready to sacrifice for God.

The following morning Abrar spoke with her neighbor Ala'a, the Sunni man who had fled eastern Baghdad with his wife during the sectarian bloodletting in the capital. *Ala'a*, she told him, *I need you to show me how people in your old neighborhood get their water.*

Water tanks are such a common sight in Baghdad, so much so that residents never give them a second glance. An estimated one in four Iraqis have no access to clean drinking water, a reality of post-Saddam life. Some of the shortage is due to broken or out-of-date infrastructure. In Baghdad, most of the water deficit is due to the lack of capacity. Since 2003, the population of the capital had increased by upward of 60 percent. Iraqis fleeing violence from other parts of the country or from other areas of the city meant that neighborhoods with tens of thousands of residents, especially in the Shiite-dominated eastern districts, found themselves with no sewage system and not enough water pipes.

As Ala'a drove through his old neighborhood, Baghdad al-Jdeidah, Abrar sat in the back seat observing the routes of the water trucks through the residential streets. Sometimes, she had Ala'a go ask people as they received drums full of potable water how often the fresh tanks were delivered. Soon, she believed she knew more about the city's potable water sources than most of the people at Baghdad University.

The Falcons remained at a cautious remove, watching and wondering what Abrar was planning. On the fifth day of surveillance, Abu Ali had a breakthrough. One of the team's most trusted assets, a senior Islamic State official in Al Qaim, made contact.

Beware the Daughter of Iraq, he told the Falcons. She has a poison and she knows how to kill.

Finally, the puzzle pieces started falling into place. Abu Ali heard from the informant how a young woman had shown up in the border town the previous year and organized a series of experiments. Feral dogs that roamed the streets died all at the same time. So did two hutches of rabbits. The Bedouin who lived just outside of town started murmuring about the witch who killed animals with a single glance. And then, suddenly, the young woman was gone. The commander of Al Qaim had sent her to Mosul. When Abu Ali heard this, he decided that they needed to close the trap without delay.

The spymaster instructed an officer from his cyber division who specialized in infiltrating jihadi chat rooms to find Bint al-Iraq online and befriend her. After learning that she had been recruiting foot soldiers, Abu Ali told his officer to volunteer for her mission. Tell her, he said, that you are the answer to her prayers.

The following day, a credulous Abrar revealed to the officer that she wanted to attack Baghdad with a deadly toxin. We have the power to kill the kuffar, the infidels, she told him. It will take some precise handling, but we can succeed.

Abu Ali was aghast as he read the chats. He thought he finally had enough evidence to get an arrest warrant. But his pragmatic side told him that they should still hold off. Were there other members of an Islamic State cell here in Baghdad waiting for Abrar to make contact? If they detained her, would someone else carry out the plot without her?

The spymaster took a deep breath and decided to wait to spring his trap against her. The Americans were unhappy with his delay. But the decision was Abu Ali's to make and his alone. He told his cyber specialist to arrange to meet Abrar and urge her to move ahead with the attack. If she has a chemical toxin, we want to make sure we get it from her, he said. Ask her to leave you a sample somewhere so we can see exactly what we are dealing with.

Abrar expressed no misgivings about the request. She promised to drop off her present the following day at a location the Falcons had selected: a dry goods store in the Karkh district near the city's central train station.

The owner of the store was related to a member of the Falcons and could be trusted to do exactly what he was told without asking difficult questions. Abu Ali wanted to make sure that no one touched Abrar's package and that it wouldn't be misplaced.

His plan kicked off the next day. One of his men staked out the street outside the dry goods shop and he was ordered to keep a low profile. Abu Ali didn't want to flood the area with too many men. Abrar, he thought, would be nervous and she was bound to feel all the extra eyes on her.

He updated the Americans on his game plan, but not the prime minister, or Baghdad Operations Command. Sharing details with too many people could ruin such a delicate mission. He wanted a better understanding of the threat before alerting the rest of Iraq's generals.

The morning of Thursday, September 29, started hot and sunny and it only got warmer as the day progressed. Abrar was up early, as always, for the sunrise call to prayer. She had not unpacked her black satchel but had stored it carefully under her bed. She had been tempted to open the canisters and unwrap the milky-colored paste, but didn't for fear of having the toxin spill or degrade.

She hadn't figured out how to use her new Islamic State volunteer who had so enthusiastically offered to help. Abrar had eight cakes of ricin. Divide each of them in half, and she and her volunteer could poison sixteen water tanks. Water was the perfect delivery system because the toxin would be ingested or absorbed through the skin. Hundreds would perish.

Abrar decided she would fully brief the volunteer only after she had a chance to size him up and determine his technical knowledge so he could be deployed effectively. In the meantime, Abrar divided her supply of ricin in half. She planned to leave a portion for him in Karkh and keep some for herself, in case this turned out to be her only chance of delivering the weapon she had so carefully created.

Inshallah, she thought, I hope everything goes well.

For the next several hours, events proceeded as she had hoped. Ala'a, her driver, dropped her off at the dry goods store and, when she went inside, the shopkeeper was amenable to her bidding. She told him that she had a bag to drop off for her cousin, who was running late, and asked if he would store it until the cousin came to collect it.

I'll pay you for your trouble, she told him. It's just that I have to get back to my family, and my home is so far away. I don't want to inconvenience my cousin.

The shopkeeper placed the black satchel behind the cash register. Don't worry, my dear, he told her. When he arrives I will give it to him.

She was in a good mood as she climbed back into Ala'a's white sedan. She then asked him if he had time to take her one more place. She didn't want to scare him but she indicated that the location would be a bit dangerous. Could you take me to Sadr City? Abrar asked.

She didn't know anything about the sprawling Shiite district, but wanted to make a similar reconnaissance to study its water distribution system as well. If the goal of a terror attack was to target as many unbelievers as possible, Sadr City was the place to do it. It was a wonder, Abrar thought at the time, that she hadn't considered targeting the neighborhood until that point.

The Falcons surveillance team lost Ala'a's car in the heavy traffic of Baghdad al-Jdeidah. At a key intersection, the Falcons gambled and decided to head north, instead of east toward Sadr City, never imagining that two Sunnis would knowingly drive into the Shiite stronghold.

A few hours later, the consequences of this mistake became dire. The forensics tests on the substance that the cyber specialist had retrieved from the Karkh shop confirmed that the Daughter of Iraq really was in possession of a deadly toxin. But by then, she had disappeared.

Abu Ali al-Basri could not remember feeling as anxious as he did that afternoon. He had no way to find his suspect and no knowledge of when she intended to launch her attack. Just around sunset, a second Falcons surveillance team assigned to watch her home reported in. Abrar and her neighbor were back in Amariyah. The fright over losing his prey sparked Abu Ali to action.

He called the Americans one more time. We're going in tonight, he told them.

That Thursday night unfolded in its usual fashion at the al-Kubaisi home. After a light supper with the family, Professor al-Kubaisi sat in the living room in his beige plush armchair and listened to the evening news programs. Um Mustafa cleaned the kitchen. She didn't know where Abrar had spent her day, but her daughter took an extra-long shower that evening after complaining of the heat and grime of the city.

Abrar was eager for her family to fall asleep so she could get online on the family computer in the living room and share details from her reconnaissance that day. She had a possible target identified in Sadr City, a main square where the United Nations distributed water each week.

It was almost midnight when the house was finally quiet. But when Abrar logged on, her usual chat rooms were empty—the aspiring jihadis who always agreed with her wildest plots weren't online. Abrar spent a few desultory minutes checking into other Islamist discussion groups but decided she hadn't the patience for theological debates. She turned off the light and went to sleep.

Minutes later, masked commandos dressed in black quietly surrounded the block where the al-Kubaisis lived. An armed counterterrorism team took positions on the neighbors' roof as two other units advanced on foot toward Abrar's front door. The attack plan was straightforward: storm the family's home, neutralize the young woman, and seize any materials that look like powder or soap. Thousands of lives were in the balance, the commander told his men.

Shortly after two A.M., Abrar was jolted from sleep, startled by the crack of a concussive grenade. Before she knew what was happening, men wearing balaclavas and gas masks pulled off her covers and yanked her from her bed. The room was pitch-black except for the flashing spots in her eyes. She couldn't see or hear anything. The commandos didn't give her a chance to dress properly, they just threw her black abaya over her shoulders and put a black bag over her head.

Men barked orders. Identify yourself, one of her captors demanded. That's when Abrar realized she wasn't dreaming. She felt cold and angry. She knew that her plans were dashed, but she answered her inquisitors with as much dignity as she could muster. Abrar bint Mohammed al-Kubaisi, she said. How dare you put your hands on me.

She knew right away what the commandos were searching for, but she couldn't understand how they had discovered her. She was supposed to be invisible, gliding like a ghost, until she launched the attack. After that, hers would be a name that no one would ever forget.

Further down the hallway, Professor al-Kubaisi and his youngest son had been forced to their knees by the men in black. The intruders spread throughout their home without saying a word. In their terror, the al-Kubaisis didn't notice that the team was collecting all the family's electronics—phones, tablets, and the desktop computer. They also took the hunting rifle that Abrar's mother had brandished when the security forces had burst through her bedroom door. The family was so overwhelmed by the home invasion that they did not notice when the masked commandos found the true object of their search: the canisters under Abrar's bed.

The leader of the unit never identified himself to the family, but as he and his team left the al-Kubaisis' broken home he curtly informed them that Abrar was being detained. You won't see her again for a very long time, he told them. He was right. It took almost six months for the al-Kubaisis to locate their daughter. By then, she had been charged with four counts of terrorism.

Chapter 20
Racing Against Time

At three P.M. on an unusually sunny December day in 2016, Harith heard the familiar melody, the same one that had summoned him at this time each week for more than a year. He pulled the vibrating black Samsung phone from his pocket and the sound of a soaring recitation of holy verse grew louder. Punctual as always, it was Abu Qaswarah from Mosul.

Salaam wa Alaikum, my sheikh, Harith greeted the Iraqi Islamic State commander. May God grant you good health.

Peace be upon you, O soldier of Islam, Abu Qaswarah replied. Praise God, I have news for you.

It was Harith's sixteenth month undercover with the Islamic State. He would never know it from the unfailingly upbeat conversations he had with Abu Qaswarah and his own cell leader, Abu Mariam, but from the time he had started his mission until the end of 2016, the once seemingly invincible caliphate had been brought to its knees, at least in conventional military terms.

Huge swathes of territory captured by the militants in the summer of 2014 had been retaken by government forces in hard-fought battles. The insurgents had been outgunned and outmaneuvered by Iraqi military generals supported by the American-led coalition's jet fighters and long-range missiles. Together, they had pounded the terror group's armories, financial assets, and strongholds. As important, Baghdad was winning the psychological war, and the delicate equilibrium between hope and despair had tipped in favor of the former, crippling the caliphate's ability to recruit and fight another day.

The al-Sudani brothers had been a secret weapon on these front lines. Harith, living in Tarmiya with the enemy, and his Falcons colleagues in Baghdad, had helped restore a semblance of calm in the Iraqi capital by disrupting thirty planned suicide bombings and seventeen vehicle bomb attacks. Rising morale had helped the fragile government coalition focus on the ground war. It had also helped public morale. By December, people across Baghdad felt the change. Munaf, his wife, and their neighbors all talked about the lack of tension in their shoulders as they pulled into traffic or walked into one of the capital's new shopping malls.

Still, Abu Qaswarah, a diehard warrior of the caliphate, exuded a heady confidence every time he spoke to Harith. His fighters may be losing ground, but his public commitment to the jihad against America and the American-backed government in Baghdad never wavered. Harith never could tell whether Abu Qaswarah harbored any doubts about the caliphate, its tactics, or the wisdom of its leaders. After all, if anyone in Mosul were foolish enough to express such reservations, his life would be forfeit.

On this December afternoon, during their weekly phone call, Abu Qaswarah described to Harith his next plan of attack against Baghdad. His orders were chilling.

Abu Suhaib, the Mosul commander told him, we have another gift to send to your aunt. We want her to receive this package soon, on the last day of the infidel's calendar.

"Package" was the Islamic State's code name for a vehicle bomb—and the target, as usual for Harith, was the Iraqi capital. But that wasn't the whole message. As Harith listened, he realized that this upcoming operation, his forty-eighth on behalf of the Islamic State, would be unusual. It was part of a far more ambitious offensive being worked up in Mosul.

Our blessed organization is preparing presents for many relatives around the world. The unbelievers will receive these gifts on their Christmas holiday. But, inshallah, these gifts will bring us joy, not them, Abu Qaswarah continued.

Harith immediately understood what the terrorist leader meant. It was only a week before the end of the year, and the Islamic State was searching for a way to make a sweeping statement to the world that the group remained relevant and had the power to disrupt life as the West knew it.

I am ready, sheikh, Harith told his commander. Tell me what I need to do.

The present is being wrapped and will be delivered to you. Be patient. A man from Anbar will be in contact and he will give it to you. And then you will need to make sure it reaches its final destination. Be ready, my brother.

I have never failed you before, my sheikh. I won't fail you now, Harith replied.

The men ended their call, and Harith immediately opened Telegram.

There, Harith read more about the contents of the package and instructions about its delivery. The proposed attack called for a farmer from Al Qaim, the Iraqi border town where the Islamic State operated bomb factories, to drive a white Kia truck to the outskirts of Baghdad. The truck bomb would be packed with six thousand pounds of explosives, a payload capable of killing and maiming thousands of Iraqis. Harith would take delivery of the truck from the farmer and, on December 31, drive it to the open-air markets of Baghdad al-Jdeidah, where he would detonate it amid the Shiite, Sunni, and Christian shoppers looking for last-minute presents for New Year's Eve.

After he finished reading his orders, Harith signed out of the account he had created as Abu Suhaib and signed back on to a second encrypted Telegram channel, the account through which he communicated with Munaf.

Brother, I have news, Harith wrote. Packages are being delivered on December 31. Not only to our family in Baghdad, but also to other relatives around the world.

Just as quickly as Harith logged in, he logged off again. He couldn't call his brother's usual contact line. He had no time and no ready excuse to go to the gas station or another spot where he could have privacy. He had to trust that Munaf would log into Telegram and see the message before the day was out.

After all, the clock was ticking. There were only four days left to thwart the Islamic State's global assault.

In Baghdad that same morning Munaf was once again stuck in one of the city's notorious traffic jams, his pet peeve. He had so much paperwork, so many operations, so many agents that he hated wasting time in the car. His normal commute from home in Sadr City across the city to the Falcons' headquarters could take as long as ninety minutes each way. So Munaf often saved himself the daily headache by spending the morning at the secure electronics surveillance center at the Iraqi Interior Ministry, located in northeastern Baghdad and a quick drive from his home.

In the byzantine ways of Iraqi bureaucracy, Munaf, like many of the Falcons officers, took advantage of the resources of friendly security agencies that their team lacked. Abu Ali's close relationship with his Interior Ministry colleagues helped foster cooperation, when other agencies, such as the mukhabarat, slammed their doors shut.

All morning long, Munaf had reviewed reports of excited online chatter monitored by Iraqi intelligence. When he opened Harith's message later that afternoon, he realized why excitement was building among the jihadis. A big operation was in the works. Harith had supplied enough information to prevent an atrocity in Baghdad, but what about the rest of the world?

Brother, tell me more about where the presents are going to be delivered? To our neighbors, or farther away? Send details when you can, Munaf replied.

He logged off and called Abu Ali al-Basri right away.

Chief, we have a situation, Munaf said. A new target in Baghdad, as well as other cities abroad. We don't have much time. The plot is set for New Year's Eve.

By the end of the day, the message had been relayed across Europe, Russia, and the UK, where intelligence agencies were already jittery about the Islamic State's international reach. Earlier that year, men who had pledged allegiance to the Islamic State had bombed the Brussels airport in Belgium; mowed down pedestrians in Nice, France; opened fire at the Pulse nightclub in Orlando, Florida; and just days before Harith had sent the encrypted message to Munaf, a man had killed shoppers at an open-air Christmas market in Berlin, Germany.

German authorities put police on high alert through the holiday week. In Russia, authorities evacuated the capital's train stations ahead of New Year's Eve, after receiving warnings that bombs had been planted there. The heightened security warnings weren't enough in Turkey. Just before midnight in Turkey's sprawling metropolis of Istanbul, a lone gunman opened fire in a packed nightclub. He killed thirty-nine people and wounded dozens more. When he was caught by Turkish police weeks later, he confessed to taking orders from an Islamic State commander in Raqqa.

At the time, neither Munaf nor Harith knew about the precautions elsewhere. They were too busy preparing for what lay ahead of them in Baghdad. Munaf had his whole team on high alert, ready to sweep into action. It went without saying that after the Karada bombing four months earlier, the Falcons could ill afford another mistake in the Iraqi capital.

Munaf expected to make contact with his brother sometime on December 30, the date on which Harith had been told to expect the explosive-laden vehicle to arrive.

It had been two weeks since Munaf had last seen his brother, and he was increasingly worried. His dreams, Harith had told Munaf, were almost always of death. Munaf didn't have a background in psychology, but he didn't need a university degree to understand the connection between those morbid dreams and the stress Harith was experiencing. But, as a brother, he had kept his word. He had not relayed this information to Abu Ali. He never told anyone about Harith's unauthorized trip to Sadr City, and the risk he took of being caught in a lie.

Harith always assured Munaf that he could handle himself. And Munaf took him at his word. His reports to the Falcons' chief, after every debriefing and successfully prevented terror attack, always indicated that Harith appeared to be faring well.

Now, as December was at a close, he sat with Harith in the Falcons' office near Baghdad International Airport on a hard-backed wooden sofa surrounded by ashtrays and paper cups of strong black tea. Harith looked more at ease than he had in the past year and a half, Munaf thought. But his face was puffy and his eyes were red. Harith told him he still wasn't sleeping well.

Harith had come into the capital expecting to take receipt of the truck bomb. But hours had gone by without any word from the farmer from Al Qaim, so Munaf had taken him back to their old office, the compound where they had worked together before Operation Lion's Den began. As word passed through the hallways that Harith was visiting, men who hadn't seen him for months stopped by to say hello and slap him on the back. The room was warm, fuggy with cigarette smoke and laughter.

Munaf saw the way his brother came to life with the attention from his colleagues, but still something was wrong. He noticed a tremor in Harith's hand as he drank his tea. Munaf had never known an alcoholic, but he had interrogated several drug addicts over the years. He noticed that Harith was exhibiting some of the same tics that men going through withdrawal suffered.

Quietly, Munaf leaned over to his brother.

I'm afraid this job is going to kill you, Harith, he said. Do you want to pull out?

Harith turned to look at him and Munaf saw a flash of anger in his eyes.

Hmaar—you donkey, Harith said. Is there someone else in this room who can take my place? If I don't do this, who will?

Munaf looked around their squad room and knew his brother was right. Since Harith's undercover training in the summer of 2015, Abu Ali had passed word that he would like to have more officers trained for undercover missions to penetrate the terror group, but no one had volunteered. Without Harith, more than three dozen bombs would have hit Baghdad, and the whole course of the war could have changed. Instead, Iraqi government forces were preparing to liberate Mosul, something they would not have considered possible just two and a half years ago.

Without Harith, hundreds of Baghdad's residents could die the following day, killed by the truck bomb that the Falcons were waiting to intercept.

Munaf didn't answer his brother. Instead, he changed the subject and asked him what he wanted for lunch.

The brothers ate Harith's favorite grilled kebab and spent the evening smoking their water pipes, waiting for the phone call from the courier. By midnight, with no updates, the two dozed off. Just before sunrise, Harith finally received the message he had been waiting for.

The visitor from Al Qaim was expected to arrive at nine A.M. at his relative's house in Adamiya, the Sunni-dominated northern Baghdad neighborhood.

Harith quickly showered and scrubbed away the smell of cigarettes and any traces of his time with the Falcons. Next, he gave himself a close shave. He left the tight black T-shirt that he had borrowed from Munaf in a clump on the floor and put on the clothes of Abu Suhaib—a clean, loose-fitting button-down shirt and baggy trousers. He emptied his pockets and went back to the duty room, where Munaf had started his briefing about the upcoming operation.

The mission was expected to be straightforward, like the previous seventeen car bombs that the Falcons had successfully prevented.

Munaf took charge of a two-car surveillance team that would monitor Harith as he took possession of the Kia truck. One car would jam all communications around the truck bomb, to prevent any remote triggering of the explosives, which was a favorite Islamic State tactic. The second car, meanwhile, would be filled with the Falcons' bomb ordnance team, the men who would defuse the explosives before Harith reached the target.

Munaf was all business. No detail was missed. No possibility left unexamined. He caught his brother's eye from across the room and nodded. Harith slipped out the door, back to his life as Abu Suhaib.

Outside, the weather was chilly and gray clouds were spitting rain on the highway. The older al-Sudani had done what his alias was expected to do. He had caught a bus to Adamiya, while the Falcons followed at a distance. As the bus lumbered over the Tigris, flowing swift and gray like molten silver, Harith flashed back to the cold panic of his nightmare, the one in which he was swept underwater and drowned.

He had told Munaf not long before how he was struggling to block those feelings and keep the doors with which he had compartmentalized his two identities closed. He willed himself to concentrate on the mission that was ahead of him and the burning question he often pondered: What would his family think if they knew what he was doing, and the sacrifice he was making for them and the nation? Would they finally look at him the way they looked at Munaf when he came home from work, with respect and pride?

As the bus drove across the overpass from the airport road to Adamiya's market street, Harith pulled his phone out of his pocket and typed out a brief message to his father. "Pray for me," it said.

Harith's rendezvous with the farmer took place at a crumbling, unpainted two-story cement building along a rutted, dirt-packed alley. Like so many homes in that part of Baghdad, it could house one large family or half a dozen. The exterior was encrusted with several layers of sand and grit. Rusting rebar wires stood up in discordant angles on the top floor, adding to the aura of neglect. Rainwater, mixed with sewage seeping out of an open pipe, formed a large puddle across the narrow alley. Harith knew he was in the right place when, through the dripping rain, he noticed an open-bed white Kia pickup truck. An elderly man with a dishwater gray beard stepped out of a doorway as Harith approached.

Peace be with you, Abu Suhaib, he said, extending his hands to the younger man.

God grant you peace, uncle, especially after your long journey. Welcome to Baghdad.

The older man was moving stiffly, his muscles cramped from the long drive. He chatted easily with Harith, telling him about the trip, the traffic, and the weather along the road. His face was tanned and raw boned, from a lifetime of outdoor work.

The farmer said he had been unexpectedly delayed when the army closed the checkpoint at the northern edge of Baghdad for the entire night. He and dozens of drivers were not told the reason for the closure and had slept in their cars. Just before dawn, a new shift came on duty at the checkpoint and let the vehicles pass.

The old man's eyes were tired, but curiously free of worry. Harith later described to his brother how astonished he was at the man's composure. He doesn't know that he has been driving a bomb, Harith thought to himself. Neither, apparently, did the soldiers who allowed the vehicle to pass toward Baghdad.

Harith pulled $300 from his pocket and handed it to the older man, the price he was promised for ferrying the truck.

Thank you for your service, Harith said. God grant you a safe journey back home.

Munaf and another man from his team watched the exchange from a distance. He knew there was little danger so far. The Islamic State wouldn't detonate their vehicle in the middle of a working-class Sunni neighborhood. They wanted to kill Shiites, or Christians, people who would be shopping ahead of the Western New Year. The younger al-Sudani was already focused on Harith's next step: the route his brother would drive the truck bomb. That was always the most dangerous juncture of these interdiction operations, the time when Harith was most vulnerable. These were the moments he needed to be as close to his brother as possible.

In the alley, Harith shook hands with the old man and sat down in the pickup. Like all the other vehicle bombs he had driven, the Kia was old and beat-up well before being repurposed as a weapon. The seat belts were broken. The radio was a simple, factory model with an AM/FM radio dial and a slot for a cassette tape. Not that Harith could have listened to music, even if he wanted to. The bomb makers had to rewire the vehicle's electronics for the detonation. Electric windows wouldn't work, nor would a radio.

Harith took a deep breath and called Abu Qaswarah in Mosul.

My sheikh, I have the present. Where do I deliver it?

You will take it to the place you chose for me weeks ago, the Mosul commander replied.

Harith understood his meaning. The confirmed target was Baghdad al-Jdeideh, New Baghdad, the district south of Sadr City that was home to two of the capital's most popular churches, a major bus station, and a cinema. It was the kind of neighborhood where middle-class families, civil servants, and teachers shopped for clothes for the children, and perhaps a toy or two at small family-run shops, before popping into one of the small restaurants for ice cream sundaes or fresh fruit juice. Now, just ahead of New Year's, shops would be packed, despite the cold rain. If the Islamic State wanted to strike terror into the hearts of the Iraqi people, then killing families at Baghdad al-Jdeideh would be a good place for it.

Harith hung up, turned on the ignition, and put the truck into gear. It lurched forward, the chassis groaning under the weight of the explosives that had been packed between the panels of the car.

For now, Harith concentrated on keeping his eyes on the road and steering clear of the cars around him that might accidentally jostle or bump him. Until Munaf could intercept him, any minor traffic accident could be enough to set the bomb off prematurely.

As he drove down Adamiya's main commercial street, past the apartments built for displaced Sunni families and the stretch of road where truffle farmers sold their delicacies each spring, Harith finally saw Munaf's black four-door sedan. As he passed, he gave his brother a thumbs-up.

Baghdad al-Jdeideh, Harith yelled out the window. I will leave the bomb by the Baghdad Cinema, near the bus station.

The Falcons' after-action report described the drama on the rain-slicked streets along the Baghdad highway that late December morning, as Harith tried to keep thoughts of death at bay.

He was safe, he kept telling himself, but as the windshield wipers swept away the drizzle, Harith remembered all the videos he had watched at the Tarmiya farmhouse of spies being killed. He saw clearly the flat-eyed stare of the men who were about to die and how they ended up with a sword to their throat or gun to the back of their head. What would happen to him if he were caught by the terror group? Could the men with whom he had been living for these long months actually turn on him, torture him, and break him? Would he, in his agony, reveal his true identity?

Distracted, Harith realized he had missed the turn-off to the Baghdad highway that would lead him south across the Tigris River through the center of the capital and east toward Baghdad al-Jdeideh. That was the route that he had been ordered to take, the direction spelled out on the Telegram message he had received from Abu Qaswarah a few days prior.

He looked out his window and saw Munaf in his chase car, giving him a thumbs-up and mouthing a question as to whether everything was okay. But just then, Harith's phone rang, something that shouldn't have happened if the Falcons' jamming device had been working properly. Out of the silence came the ominous sound of the Koranic prayer. Mosul was calling again.

Salaam wa Alaikum, Abu Suhaib, the commander said. His normally terse voice was lined with an extra coat of steel.

Brother, tell me where you are, right now. And I mean right now.

Harith's mouth went dry. Did the terror group know he had made a mistake?

Harith did what came naturally after sixteen months of life undercover—he lied.

My sheikh. I'm on the Baghdad highway, near Dora. All is going according to plan, he said.

La, my son. No. You are not where you are supposed to be. You are on the ring road.

Harith felt his blood freeze. Looking around, he was certain that he would see someone from his Tarmiya cell on the road nearby. But there was no one except Munaf and the bomb squad flanking him in their vehicles.

Harith didn't understand how, for the second time in a matter of months, he had been caught in a compromising situation. He felt in the pit of his stomach that something was terribly wrong. But now was not the time to dwell on it, let alone panic, certainly not while sitting amid hundreds of pounds of military-grade explosives, fertilizer, and ball bearings.

My sheikh, you have nothing to worry about, Harith stammered into the phone. The rain is falling here and the roads are very tight with traffic. I couldn't make the turn for the exit because of the traffic.

Harith leaned into the steering wheel, putting his right elbow on it to keep it steady while holding the phone to his ear. With his other arm, he started gesticulating wildly out the window, hoping Munaf would notice. They needed to be closer because the jamming device wasn't working.

I am fulfilling my mission, Abu Qaswarah. I swear on my life, I will deliver the package as ordered. I am the loyal soldier to the caliph.

The commander still had steel in his tone.

God is great, Abu Suhaib. May he bless your journey.

Just then, Munaf's chase car pulled up directly in front of the Kia, and Harith's phone went dead. He didn't know whether Abu Qaswarah hung up, or the jamming device started working. He threw the handset back on the passenger seat and put both hands on the wheel. The last thing he wanted was to die inside this car, scared and soaked from the rain.

Munaf felt his own stab of panic. What was his brother doing on the phone? He screamed at his driver to stay as close to the Kia as possible and radioed the second car to ride directly behind Harith.

There was no time to stop, so Munaf gestured to his brother that he would take the lead. The road they were on was about to merge into Baghdad's northern ring road, which would bypass the center of the city and lead them eastward past Sadr City to Baghdad al-Jdeideh and the Falcons' safe house where the explosives unit could dismantle the bomb. The highway had no traffic stops and it was far from densely populated neighborhoods.

Harith kept gesturing at Munaf. This was not the route that Abu Qaswarah directed him to take. But Munaf wouldn't slow down, and the phones were dead. So Harith resigned himself to fate.

As the three-car convoy passed Sadr City, Munaf thought of his family just blocks away and said a small prayer. They were only a few short miles from the safe house. Let us all finish this safely, he prayed.

Minutes later, the Falcons exited the highway and entered an industrial wasteland of empty sandlots overrun with thorny scrub brush and piles of construction waste. The street was almost empty. The rain let up and Munaf felt a restored sense of calm as his driver turned left at a gap in a cinder block fence and pulled into a large acre-sized lot. He stopped the car in a cement driveway in front of an abandoned, half-built home.

The four men inside the second car started to dismantle the bomb. With the efficiency of a Formula One pit crew, they helped Harith from the truck and then pulled apart the Kia's side paneling, revealing two layers of tightly wrapped brick-size bundles. Two minutes later, they updated Munaf: They had retrieved a quarter ton of military grade C-4 explosive. Behind the seats in the truck's cab, they had found bags of ammonium nitrate fertilizer and ball bearings. Those packages were being stacked in separate piles, out of the rain. Behind the dashboard was a spiderweb of jerry-rigged wiring that connected the detonator to explosives.

Harith paced back and forth in front of Munaf's car. He said he wanted to smoke and asked Munaf's driver for a cigarette, something Munaf had never seen his brother do before. He exhaled a deep pull of smoke and explained, his voice quivering, just what had gone wrong.

They're watching me. I don't know how, but they are, Harith told his brother.

Munaf struggled to hide his worry. Being caught in a second lie was bad. But watching Harith's self-control slip was even worse. He needed to help his brother restore his calm, and do it quickly. They only had a couple of minutes to get him and the Kia back on the road. Munaf poured Harith a small cup of tea from the thermos that his driver always kept in his trunk and sat him in his Toyota in front of the heater, keeping an eye on his watch and the piles of explosives growing on the cement driveway.

Harith, take it easy and drink your tea. There is no one watching you. No one knows where we are, Munaf said.

Harith scanned the horizon. No cars had stopped along the road. There was no bystander within sight of the safe house.

An excited yell from one of the bomb-disposal specialists drew the brothers' attention back to the Kia. From underneath the roof paneling of the pickup, the sapper had found an anti-tank mine. He and another officer were carefully moving the twenty-pound munition to safe ground.

Another member of the bomb squad shouted for Munaf. He was holding up a small Samsung phone that he had found in the glove compartment. The wiring attached to the handset led to what had originally been a cigarette lighter inside the truck. Munaf asked his brother if he had ever seen the phone before. Harith shook his head. The technician guessed that it had been intended as some sort of a secondary detonator for the bomb. He capped the wires and put it back where he had found it.

Five minutes had passed. At most, they had another five minutes before they needed to be back on the road, so that no further suspicion could fall on Harith. The bomb squad scrambled to fill the truck paneling with sandbags to make it appear that the truck had not been tampered with, just another precaution the Falcons took to maintain Harith's cover. A couple of minutes later, the vehicle was fully reassembled, transformed from a lethal improvised explosive device back into a harmless old clunker.

Harith finished his tea and slid back into the driver's seat. He wasn't going to leave the job unfinished.

The rain stopped as Harith moved south in the early-afternoon traffic on Darwish Street, the popular shopping district in Baghdad al-Jdeideh. He crawled along as taxis and minibuses choked the boulevard. Iraq has long considered New Year's Day a national holiday, not only because the country is home to one of the oldest and most vibrant Christian communities in the Middle East, but also because the government wanted to remind its allies that Iraq shared their Western values. It wasn't like the fundamentalist countries to its east and west—Iran and Saudi Arabia—that shunned Western ways.

Harith drove past the kind of mom-and-pop stores that cater to the district's middle-class shoppers, including toy stores specializing in bikes with training wheels and life-size dolls made in China. Mothers wrapped in headscarves and winter coats pawed through the displays of children's clothes for sale by street vendors in market stalls. The Baghdad Castle restaurant displayed a Christmas tree in its window and families filled its red booths to eat hamburgers and drink pomegranate juice.

At the intersection, just past the post office, Harith turned right onto Ragheer Street, and headed toward the city bus terminal, the local police headquarters, and the Father of the Apostles Cathedral. Abu Qaswarah had told Harith it was up to him to decide which of these landmarks to target.

Harith circled the neighborhood twice, enough time for him to scout the layout of the street but not long enough for him to attract any undue attention. The last thing he wanted was a bored traffic cop to stop and harass him. Maintaining his cover story with the Islamic State depended on following the mission through to the end. That meant Harith needed to park the vehicle, leave the scene unnoticed, and then let the Falcons finish their task, which was to initiate a fake explosion so that the Islamic State would believe that Abu Suhaib had orchestrated another successful attack.

Harith could see immediately that the Christian church was the most daunting target. Iraqi police had surrounded the building with concrete blast walls because of the holiday, and a special detachment of Interior Ministry police patrolled the church perimeter. The bus station, meanwhile, was crowded with travelers. But it had challenges of its own. The driveways in and out were jammed with grimy vehicles and there was no room to park the Kia. That left the traffic police headquarters, located next to the cinema, one of the other targets that had met the terror group's approval.

At around five P.M., Harith parked the truck in front of the main entrance and melted into the crowds that were heading, shopping bags in hand, for the bus station. The street was already dark and a bitter, chilling wind had whipped up. He was ready to get this over with. Harith texted Munaf when he was well away from the vehicle and told him he was ready to strike.

Munaf had watched his brother circle the neighborhood streets and finally park the truck. He and the other Falcons were on the other side of the intersection, far enough away from the controlled explosion they were going to trigger, but near enough to Harith in case anything went wrong.

May God protect us, Munaf texted back. Let's go.

Harith dialed the code on his phone that would set off the detonator in the truck. Immediately, he heard a loud boom, as the Kia lifted into the air, propelled by compressed air and sandbags instead of deadly projectiles and explosive material.

Pedestrians screamed in fear at the sound of the boom. The force of the blast set off car alarms, adding to the volume and hysteria in the street. Then, Munaf heard a loud whooshing noise. A truck of compressed cooking gas canisters had been driving by right as Harith set off the detonator, causing a secondary explosion and starting a fire at the three-story shopping mall across the street from the police building. Within minutes the whole building was aflame. Dozens of fire trucks and civil defense vehicles roared to the scene, while shoppers scattered as fast as they could.

In the mayhem, Harith took the opportunity to film himself in front of the growing flames. He sent the image to Abu Qaswarah and felt a rush of joy.

Oh sheikh, you see what your loyal soldier has accomplished today? Harith texted his commander.

The fires of the righteous are killing the infidels in Baghdad. Those who worship the false gods are paying for their sins.

Later that night, after a lengthy debriefing, Munaf sensed that his brother was more relaxed. From the Falcons' perspective, it had been a good day's work. From Mosul's point of view, the Islamic State could boast another successful mission.

On television, Iraqi media reported that sixteen Iraqis had died in the truck bombing in Baghdad al-Jdeideh and scores had been wounded. In fact, no one had died. As usual, Iraqi journalists had accepted what the authorities had told them. In the end, the only person who suffered any losses was the owner of the shopping center, but Munaf hoped he had insurance to reimburse him for the damages their operation caused.

Harith leaned back on the tea-stained sofa in the duty room, laced his hands behind his head, and looked at his younger brother.

We saved Iraq tonight, he told Munaf.

They both knew that was true. What they didn't know was that Harith's cover had been blown. By the time the Falcons found out, it was too late.

Chapter 21
Stretched to the Breaking Point

Firefighters had not yet fully doused the flames of the fire in Jdeidah when Munaf made an urgent phone call. Boss, we need to pull our man from the lion's den, the younger al-Sudani told Abu Ali al-Basri as he watched the emergency crews finish their work, the streets drenched with water.

The pressure has finally gotten to him, he told the Falcons' chief.

What happened, my son? What went wrong? Abu Ali asked.

Munaf couldn't find the right words. The sun had set on 2016 and the day had ended in another successful mission. Something was wrong with his brother, but Munaf didn't have one specific thing that he could point to or tell their spymaster. It was rather a collection of small things, starting from the time he had tried to sneak back home at the end of the summer. The recklessness and paranoia, but also the two threatening phone calls from Mosul. Abu Ali didn't know about the unauthorized trip to Sadr City, the mistake that started it all, and Munaf wasn't going to tell him about it now.

What the younger al-Sudani knew was that there had been only one other time in his thirty-one years that his brother had been as nervous as he had been during that day's operation. That other instance had marked his life forever, the day Harith had to tell his father that he had been kicked out of university. All the al-Sudani brothers knew back then was that Harith had a good reason for his dread. Their father would likely beat him to within an inch of his life for causing such shame to the family. This time, the stakes were even higher. The penalty for failure would be much worse than a whipping. If Harith's suspicions were right, if the Islamic State had somehow followed him, or had found them out, then his punishment would be torture and death.

Munaf didn't want that to happen. He knew, like all the Falcons did, that Abu Ali was a deliberate leader. He quietly gathered all relevant facts, listened to his subordinates, thought through the consequences of actions, and made a decision. Some colleagues from different services considered this trait a sign of weakness. Those commanders were also the men who had some of the highest casualty rates in Iraq's security forces. The Falcons trusted their leader to make the right decisions to keep both them and the nation safe.

Boss, we have to find a way to give him a couple days of rest, Munaf said. He shouldn't go back to Tarmiya right now.

Munaf and Abu Ali spoke with Harith after the debriefing on the New Year's Eve bomb. It was late in the evening, and it had been a long day already. But when he saw how haggard his undercover officer looked, Abu Ali began to comprehend what Munaf had been trying to tell him. The shadows under Harith's eyes and the way he hunched his shoulders showed how much the stress had been affecting him.

My son, Abu Ali told Harith as he welcomed him into his office. You don't look like you've slept in days. You need a chance to freshen yourself up before going back to Tarmiya. Tell your cell there that you've been delayed a day, the checkpoints have blocked the road. Go home for a night. Go home to your wife and your family.

The new year was just a couple hours old when Munaf drove home with Harith to Sadr City. Streetlights were twinkling on the crosstown highway, the same one Harith had driven on earlier that day with the truck bomb. The streets were largely quiet in the early-morning hours, except for a few revelers. Munaf shifted gears as he passed some slower traffic, a white Kia minivan full of passengers, and glanced at Harith. His brother was quiet, and seemed out of sorts, like he was floating in time. It was an eerie feeling, to be reminded of their old life, when the brothers used to commute together to the Falcons' headquarters and Harith transformed into Abu Suhaib, the jihadi who wanted to kill civilians.

When the brothers arrived, the family was starting to stir for the dawn prayers. When Um Harith saw her sons, she yelled in delight and soon the living room was filled with three generations of al-Sudanis. Everyone greeted Munaf and Harith with happy kisses before the brothers retreated with their father to the majlis, the room reserved for guests. Nothing had changed since Harith had been away, not the spring-green-colored walls, the whirring ceiling fan, or the delicate inlaid wooden coffee tables on which their father's friends would crack pistachios and rest their teacups while listening to him recite poetry or discuss the news of the day.

When he was little, Harith had been the boy pressed into service for the adults, cleaning up their nutshells and filling their teacups over and over again for hours. Now, he was being treated like the honored guest, and, as the way of Sadr City, it was a new generation of sons, including his eldest, Muamal, who were serving him.

Harith brought the chatter to a halt by asking his son to stop serving and sit down, except the boy he pointed to was his nephew, not his son. The family erupted in laughter. The boys looked alike and they thought Harith had made a joke, confusing Muamal with his cousin.

But Munaf could see what they couldn't. Harith wasn't trying to be funny. He was close to a breakdown. The cacophony of babies crying and women chattering in the kitchen one room over, noises Munaf considered soothing, seemed to be too much for Harith. He was jittery, shaking his leg perched over one knee. With Muamal by his side, he kept saying how shocked he was at how much his son, now fourteen years old, had grown.

Munaf could see the confusion in his brother's eyes. For all the months that Harith had been undercover, he had held the memory of his family close, a balm against the pressure and the stress. He had, after all, decided to undertake the mission because of the shock and fear of losing his son in a suicide attack, and the possibility that that could happen to anyone's son in Iraq. But to realize that you could not recognize your own son in your own home? That was its own kind of stress after a year and a half of the extreme pressures he had lived under.

Their father didn't seem to notice anything was amiss. While asking his sons about their work, Abu Harith used the sort of tone reserved for honored guests. In a time of war, he knew that Harith's long hours on the job meant he must be involved in something important, a feeling that Munaf had confirmed every time their father or Raghad had complained that he had been neglecting his duties at home.

Harith had little to say. He barely seemed to register the deferential attitude their father had taken with his sons. Munaf decided that he needed to excuse both of them, so Harith could go upstairs to the relative quiet of his apartment and sleep.

Raghad had prepared their bed for him. She had learned in the fifteen years of marriage that her husband would grow cold if she approached him straightaway with her problems, and there was no use trying that early morning. She could see he could barely keep his eyes open. She told Muamal to stay downstairs with his cousins and she kept their two daughters busy with chores in the second of their two rooms. She would find a way to approach Harith later that day.

Her husband slept more than twelve hours, and the sun was setting when he finally got out of bed. And even then he could barely lift his head to eat. The chatter of the children, even the pleading of his youngest daughter to play, couldn't rouse him. Um Harith was instantly worried, thinking her son must be terribly sick not to eat the kouzi, the slow-cooked stew of lamb's meat and rice, that she had made specially for her officer sons, home from the war.

Munaf told her not to nag, and he asked Raghad to leave him be. Harith wasn't sick, he just needed more sleep. His children were told to stay out of the apartment, and Harith asked for nothing but solitude.

He deserves this, Munaf told them. I swear to God you will never know how much he deserves this.

The following day, Harith woke with the morning call to prayer, but with the same unsettling feeling of dislocation. The man's voice blanketing the street with his melodic reminder to the faithful was the voice of his childhood, not the hacking sounds of Abu Mariam in Tarmiya shaking each of the jihadis awake and to their duties before God.

He went downstairs to the sink just off the kitchen where he knew his father would be performing his own ritual ablutions. The rest of the house was silent except for Um Harith, who was getting the gas lit on the stove for the morning tea.

Salaam wa Alaikum, Harith greeted his parents and joined his father at the washbasin.

His father watched as Harith bent over to pour water on his hands and feet, using the motions that he had taught him when he was a little boy. Abu Harith felt there was a strength in his son that he had never noticed before. Where before he had observed a strong will and stubbornness, now he felt resolve. Harith had found a purpose in this new job of his, his father thought. But there was something troubling him, something stormy in his dark brown eyes.

Harith walked to the cabinet where his father kept his prayer mat. He positioned his father's carpet on the floor facing east, in the direction of Mecca, with his own to the right.

As his father knelt down he tucked his long, loose shirt under his knees and turned to his son.

What did you mean when you texted me the other day? Abu Harith asked his son. What was I supposed to pray about for you?

Harith looked at his father for a long time in silence. Abu Harith could see from the set of his shoulders that he was struggling over what to say. The older man had no idea what could be weighing on his son's mind. Surely there was no problem at work. If there was something, some political struggle or personality conflict, Munaf would have told him well before now. There was something deeper going on. But he wouldn't pry. Munaf had told them to leave Harith be and that's what he would do.

In the end, Harith remained silent and the two men bowed their heads to pray.

The news broke as Um Harith was laying out the breakfast dishes on the sofra. Munaf came running down the stairs from his own apartment.

A suicide bomber, he shouted when he saw Harith sitting next to their father, already sipping their tea. Another bomb in Baghdad!

Harith visibly shrank against the floor cushions and his face fell. A mere thirty-six hours of rest between his operation in Baghdad al-Jdeideh and now this second attack he hadn't known about. Why hadn't Abu Qaswarah told him that there was a second present being planned for the capital? It was like they didn't trust him anymore.

The dead? he asked Munaf. How many dead?

It just happened a few minutes ago, in the used car market, his brother replied. I'm sure that it will be bad. Harith stood up. Let's go, he said to his brother. Drink your tea and let's go.

Go where? Munaf asked, perplexed about the drastic change in his brother. Harith had been like a zombie for the last day and a half, but suddenly it was like he had injected a drug in his veins. He couldn't sit still.

Back to work, Harith said. I need to get back to work.

He pushed his way past his brother and out of the family room. He bounded up the stairs back to his family's own apartment. Raghad heard the commotion he was making in the bedroom, opening and shutting their cabinet doors and gathering his clothes. She came to the doorway to make sense of what was going on.

Harith, why are you making a mess? What do you need? she asked. She couldn't understand her husband's antics. He came home like he was dead to the world, but now he was manic. He didn't seem to hear what she said.

Are you leaving again, Harith? Where are you going? You can't go just like this. You need to stay here and help me, help our children!

Harith paid no attention to his wife. He pulled out an old sports bag and started shoving some clothes inside.

Harith, listen to me, Raghad pleaded and yanked at his arm, trying to force him to look at her.

What kind of husband are you who ignores his wife's needs? she screamed. What kind of man neglects his children like you do?

Raghad could see from his face that her words had acted like a bullet. Without knowing how, she had scored some kind of direct hit.

Harith hurled the bag across the room and shouted at her. You have no idea what I am doing on behalf of our children. Have no idea what sacrifice I'm making for them.

Raghad was perplexed. All she knew was that for months her own nerves had worn so thin she was close to the breaking point. The worries and work of a single parent were too much to bear. She had no idea how to raise a son, and she didn't like having to ask another al-Sudani to help her. That was what a husband, even her husband, was supposed to be for.

She remembered the lighthearted laughter of two nights earlier when her husband had failed to recognize his own son. Without thinking, she threw the joke back at Harith, only this time she said it with barbs in her voice.

Sacrifice? You wouldn't know your own son on the street to save him if he was in danger, she said.

The anger came from nowhere. Harith grabbed her around her arms and shoved her against the wall. She screamed when she saw Harith's fist coming for her face.

At the last minute, he redirected his fury. Instead of hitting her, he punched the mirror on the wall.

The glass shattered into a thousand pieces and blood seeped from his hand.

I'm going back to work, Harith said.

Raghad felt her face redden with shame. The whole house would have heard them arguing and the loud crash of the mirror breaking.

Munaf had asked her to leave him alone, but she hadn't. She couldn't. She had too many things she needed to get off her chest. And surely the family would think the fight had been her fault, not Harith's.

Raghad got down on the floor to start picking up the tiny shards of glass. She tried to harden her heart and hold back her tears. Better for the children not to see her cry.

Ten days later, on January 12, Harith made contact with Munaf from Tarmiya.

Abu Qaswarah has a new present for me to deliver, he told his brother.

Munaf could hear the excitement back in his brother's voice, the old purposefulness that he remembered from his brother's first days undercover.

But he couldn't believe what Harith told him next.

I don't have many details to provide now, he said.

The cell is changing its procedures. Instead of meeting the vehicle in Baghdad, they are sending me to another farmhouse here in Tarmiya.

Munaf was instantly alert. For sixteen months the Islamic State had organized its operations in the same way. Why change things now?

What new farmhouse are you going to? Munaf asked.

Harith described a place closer to Fallujah, somewhere more remote. Munaf didn't recognize the name of the road, and he knew that it would be difficult to organize reconnaissance now, at this late stage, without anyone from the Tarmiya cell knowing.

The younger al-Sudani didn't like the plan. He didn't like it one bit. He would lose all contact with his undercover officer, and worse, Harith didn't seem to understand how reckless it was. Wasn't it just a little more than a week ago that he was scared he was under surveillance, that his cover had been blown?

Brother, I beg you. Think about this. You should get out of there. Get out now. This could be a trap, Munaf said.

Harith didn't hesitate. He didn't even consider what Munaf was saying.

The one time I decided to quit, the bomb made it through, he told his handler. I won't allow another one to reach Baghdad. God willing.

Munaf agonized about what to do next. Where did his duty as a brother and his responsibility as a security officer start and stop? Harith was their best chance to stop another suicide attack. But it was clear to Munaf that Harith wasn't thinking clearly. He had stopped calculating risks.

The younger al-Sudani exhorted his brother once again to pull out, and again Harith refused.

Munaf decided to find Abu Ali. He raced across the lawn at the Falcons' compound that separates his unit's offices from the commander's. Abu Ali, however, wasn't there. His adjutant couldn't say when he would be back. Munaf needed a second opinion. He wanted someone to help him shoulder the responsibility. His brother's life shouldn't be in his hands alone. Munaf spent a nervous three hours waiting for his commander. He smoked his water pipe and went through the scenarios one by one, each option worse than the next. Harith's work was more critical than ever. The battle against the Islamic State was concentrated in Mosul, in the north, and Iraq's armed forces were bogged down in street-by-street fighting. The extremists were expected to lash out where they could, and Baghdad was always the weakest link in Iraq's armor.

But as Munaf sat on the saggy brown sofa of the duty room, he admitted to himself that he was also afraid.

Losing Harith to an Islamic State trap would be worse than the deaths of unknown victims of a bombing. His family wouldn't forgive him if they knew Munaf had been in a position to save Harith and hadn't done so.

When he saw Abu Ali's Land Cruiser pull into his parking place, Munaf jogged back over to his office. He laid out the dilemma in purely operational terms.

Boss, I don't think our man is thinking clearly anymore. He doesn't understand the risk he is facing.

Abu Ali saw the pain in his officer's eyes. This was not a decision for a brother to make. He knew that. Just as he knew that Harith had understood that he might die for his country when he volunteered for the mission.

Munaf, he has made his decision. We should respect his choice, Abu Ali said. He more than any of us knows the people he is dealing with.

The next time Munaf saw his brother was six months later. It was June 2017.

Iraq's domestic intelligence in Salahuddin Province had captured a terror suspect near the shrine city of Samarra. While downloading the contents of his phone in their search for incriminating evidence, the officer in charge found a five-minute video stored on the handset. The clip showed a man with thick, dark black hair and a goatee kneeling on a sandy patch of land in a grove of palm trees. He was wearing a black Helly Hansen warm-up suit, his hands were bound by white tape, and he was shivering from the cold—and perhaps also from terror.

The intelligence officer, Major Ali Zaffrani, had grown up in Sadr City and recognized the prisoner. It was Harith al-Sudani. He had no idea how Harith had gotten himself caught up with the Islamic State, but he knew a man in distress when he saw one. He scrolled through his phone and found one of his relatives from Baghdad who knew the al-Sudani family. He left a message that Munaf should call him right away.

The two intelligence officers had a short phone conversation. Major Zaffrani didn't go into specifics with Munaf over the phone. He told him that from what they could tell, the video clip had been filmed months ago, in the winter.

I didn't know that your brother had run into any trouble, the major told Munaf.

No one knew, Munaf replied. He was on a secret mission.

The major sent the video to Munaf straightaway. It arrived in his WhatsApp messages with a cheery upbeat ding.

Munaf couldn't bring himself to open the file. He was stunned by the news. In the weeks after his brother disappeared, he had given up hope of ever finding Harith alive, or of ever knowing what had happened to him. Now that they had found a rare piece of evidence, he felt a raw burning in his stomach.

He twirled his phone on his desk several times. Munaf had heard stories of men who had been conscripted to fight for Saddam during the bloody Iran-Iraq War of the 1980s. Many had stories about being forced by their commanders to walk through desert they all knew had been seeded with land mines, instead of marching a few miles east or west to get to their destination. When the soldiers balked, the officers threatened to shoot them if they didn't obey. Munaf understood their dread as he stared at his phone. A part of him would die inside when he watched the video. He was sure of it.

He couldn't bring himself to press play. He just couldn't do it. So he got up from his desk and walked across the Falcons compound to Abu Ali's staff rooms. He told the adjutant that he had news about Captain al-Sudani.

The words caused Abu Ali to drop what he was doing and bring Munaf into his office straightaway. The two men watched a film that devastated them both.

From the first second, it was clear that whoever the filmmaker was, he went about his work without compunction. The opening scene was a close-up of Harith, his body prone on the ground, his arms bound behind him. The ground was sandy and dry and rows of large date palm trees spread out behind the eldest al-Sudani. Someone's legs appeared from off-screen, kicking him in an attempt to force him to sit upright. Clearly, he was someone's captive and Munaf could see his brother was in great pain just trying to hold himself upright.

Off camera, a man with an accent from western Iraq was barking curses at the kneeling man.

Ya kalb. Ya Mukhbir—You dog. You informer! Tell us your real name!

Munaf immediately identified the man who was speaking. After all, he had been listening to recordings of that voice for months, from the tap that they had put on Harith's Islamic State phone. It was Abu Mariam, his brother's cell leader and the man Harith had lived with for sixteen months.

If you love God then tell us, his true servants, who you really are, Abu Mariam continued.

Harith tried to sit up straighter on the hard ground. His voice was weak and halting and his nose appeared to be broken, for he was having trouble breathing in the bitter cold.

My name is Wissam Falah Daoud, like I have always told you. I am from Doura in Baghdad.

The metadata from the video indicated that the clip had been filmed two days after Harith had disappeared. It would have been an excruciating forty-eight hours, Munaf thought. He had dark thoughts about what the Islamic State men would have done to his brother over those two days. He tried to imagine the strength of will it had taken for his brother to have seemingly kept his cover story intact for this long.

We know you are kha'in—a traitor. Just tell us how they recruited you.

Harith's torso doubled over, consumed by a racking cough. He grimaced in pain and started to speak with resignation in his voice.

Two months ago I was arrested by Iraqi intelligence while I was in Baghdad, he said. They told me that I could escape jail if I would cooperate with them.

Munaf and Abu Ali could see Harith struggling to stick to a false story, and they were amazed that, through his obvious pain, he had remembered the tenets of his training. Keep your explanations close to the truth. That way it will be harder for your enemy to spot a lie.

The Iraqis told me that I could save myself and my family from their prisons, Harith continued. For the good of my family, I agreed to work with them.

All they wanted me to do was tell them when I received a bomb so they could replace the real explosives with fake materials.

There was a period of silence on the video. The only sound was the wind in the tall palm trees above Harith's head and his wheezing.

Off camera, it seemed that Abu Mariam was speaking to someone else, but the sound was too muffled to be able to make out the words.

The cell leader then turned back to his prisoner.

Zein—Okay, Abu Mariam said next. It sounded like he was running through a checklist of questions.

How many times did you inform to the rafidah? the cell commander asked.

It was one time. If was after Haji Abu Qaswarah told me about the truck to pick up in Adamiyah. The last operation.

When I picked up the vehicle, Harith explained to his captors, I called the intelligence and told them I had a car bomb.

Munaf was instantly transported back to New Year's Eve and the look of fear in his brother's eyes. He had been worried he was being watched. Munaf couldn't see then how it could be possible. In a flash he remembered the one thing different about that operation.

The phone call from Mosul and the extra cell phone that the team had found in the Kia. Was that how the Islamic State knew who his brother really was? Was that how his cover had been blown?

Harith continued his tortured explanation about how Iraqi intelligence had foiled the New Year's attack by removing the explosives from the Kia and organizing a fake bombing.

Munaf turned to Abu Ali and saw the look of concern on his face. Harith's confession, they both knew, would have sealed his fate.

So the news about the bomb was false? The bomb didn't kill anyone? Abu Mariam kept hounding his captive.

Harith, by this point, had bowed his head. It appeared that he had no more energy to keep looking up at his captor.

No, he said. The bomb never happened.

Abu Suhaib, let me play something for you now, Abu Mariam said. Something that should sound familiar.

Munaf heard something that made his skin crawl. Off camera, someone started playing another recording—it was his own voice giving orders to the Falcons' bomb disposal team.

The video clip cut out. Harith never looked up again.

Chapter 22
Unraveling

In mid-January 2017, when Abu Ali heard that Harith was missing, he had the sensation of falling down a dark hole, the same pit of emptiness where he had found himself the night his father was taken away.

Even though he was still a boy at the time, the terror in his father's eyes and the casual brutality of the mukhabarat made it clear that he would never see his father again. But that didn't stop him from playing his part in the ghoulish pantomime of life in Iraq, the days that he spent escorting his mother from jail to jail across Basra, clinging to hope.

If he had been unmasked, Harith would have no chance of survival. Abu Ali knew it and the handful of Falcons who had been read in on the mission knew it as well. But that didn't stop Munaf from desperately acting out the same role that Abu Ali had played so many years ago. He had called Abu Ali on January 14, the afternoon that Harith had gone to his rendezvous at the second farmhouse. Three hours had gone by, then four. The Falcons' reconnaissance team had no sighting of Harith, and they couldn't conduct a drive-by of the farmhouse or the road outside. It was too remote, too risky.

When the sun went down, and still there had been no word from him, Abu Ali already had accepted that the worst had happened.

Munaf, however, refused to believe what everyone else on his team knew to be true. Abu Ali had answered at least a dozen phone calls from the younger al-Sudani brother through the afternoon. It was no use trying to reason with him. It was much too early for that. The young officer's emotions had made him demanding, forgetful of the protocol of rank.

We must rescue him, Munaf repeated over and over again. We sent him there. We allowed him to go.

All of that was true, Abu Ali thought to himself. But none of it mattered. That evening he walked into the Baghdad Operations Command, past the long hallway covered in eight-foot-tall topographical maps of the Iraqi battles against the Islamic State, the color-coded first draft of history of this war. The names written in Arabic calligraphy across the tops of each hand-drawn canvas became vivid reminders of the sacrifice that thousands of men had made to oust the enemy from Iraqi land. Samarra, Fallujah, Hawijah, and now Mosul, the Islamic State's declared capital. Somehow, Abu Ali would have to convince these military commanders that they should risk their own men's lives for a rescue mission for one of his own. Why should they? These generals were losing dozens of men each day as they slogged through one of the most intense urban battles since World War II.

On January 13, the day before Harith had gone to his last meeting with the jihadis, Iraqi forces had retaken control of Mosul University. The laboratories there had been transformed into chemical weapons research facilities for the invaders, a place where Islamic State scientists had perfected new ways to kill.

Abu Ali understood the rules of warfare enough to know that he had little chance of breaking the intensity of concentration on Mosul and conveying to the generals how important his lost officer had been to them and their war effort. The chain of command had no idea that a spy had been inside enemy lines, and that was Abu Ali's biggest handicap now. He couldn't tell them about his secret weapon. For starters, he wasn't formally part of the chain of command. He answered to the prime minister, not to them. Second, and most important, if there was even a small percent of a chance that Harith was still alive, Abu Ali wouldn't be the one to blow his cover, risk a leak, and have that be the reason for his death.

Instead, Abu Ali resorted to the kind of liaison work that Iraqis know best. He reached out to individual contacts that he knew and trusted, men in a wide array of responsibilities and departments, from military intelligence to domestic intelligence and the counterterrorism force. They could help him lobby for a rescue team to be deployed.

It took three days, but he finally did succeed. On January 17, a platoon of sixteen men attacked the farmhouse that had been Harith's last known position. The team killed five suicide bombers that had been left to ambush them, and one Iraqi officer died. No one was captured alive so no one could be interrogated. They came away with no intelligence about what had happened to Harith.

More fruitful, however, was the favor Abu Ali had asked of an officer in military intelligence. The Falcons had the cell phone contacts for Abu Mariam as well as Abu Qaswarah. Both commanders had gone dark. But with the help of foreign friends, perhaps something could be salvaged from their online communications, something that could hold a clue?

For weeks Abu Ali assembled unsatisfying amounts of information, but it wasn't until June, after watching Harith's interrogation video, that the Falcons could start piecing together exactly what had happened to his officer who had helped stop four dozen suicide bombers and attacks on Baghdad.

According to Abu Mariam's farmhand, who was arrested months later, the Islamic State cell captured Harith the moment he stepped into the unknown farmhouse on January 14.

As Munaf had feared, the ruse had been a trap. The Islamic State knew that Harith was leading some kind of double life. They had placed two listening devices in the Kia truck that Harith drove on New Year's Eve. Abu Qaswarah's phone call as Harith drove through Baghdad confirmed what they already suspected: Abu Suhaib had betrayed them.

Abu Qaswarah had ordered the men in Tarmiya to kill Harith on the spot. The Mosul commander had no patience for a trial. He didn't need any more evidence than what the listening devices had already revealed.

But Abu Mariam, the farmer turned jihadi, couldn't bring himself to do it. He had lived with Harith for sixteen months and thought he had understood him, a man younger than himself but not so different in many important ways. Abu Mariam was a man of resolve, of honor, someone whose grievance was with the Shiite government, and the degrading existence that all Iraqi Sunnis had been forced to endure. He wasn't a religious extremist, not like the radicals in Mosul. Abu Mariam couldn't admit that he had been fooled by the Iraqi government intelligence agency. If he did, his own life could be forfeit. For now, the leaders in Mosul were trying to survive the onslaught by the Iraqi security forces against their city. But at some point in the future, they would have time to reflect on the defeat and hold men responsible for it.

From that perspective, Abu Mariam had no choice. He had to protect himself from any accusation of abetting a traitor. So he took Harith into custody and told the rest of the cell that their brother had betrayed them. Their anger could accomplish what he couldn't bring himself to do: punish the man who had made a fool of him.

The next two days were filled with corporal punishment, primal anger unleashed on Harith's body by the men with whom he had spent countless nights eating and sleeping. Abu Mariam had no doctor to examine the pain that had been inflicted. But it was easy to guess what the pummeling wrought by ten men could do to a body, as well as two nights of sleeping naked in the bitter cold.

Harith's confession came on the second day of captivity, on January 16. The video Munaf and Abu Ali saw had been filmed to please Mosul. He needed a record to send Abu Qaswarah so that the Islamic State's bureaucrats would be able to render their own verdict in Harith's case. There was no question that he would die. The only unknown would be where and when.

After sending the video to Mosul, Abu Qaswarah and Abu Mariam had several conversations about the traitor. Abu Mariam told Mosul that their whole cell had been compromised. In reply, Abu Qaswarah ordered him to take his men and retreat farther behind Islamic State lines to safety. Their survival was also in question.

Before sundown on January 16, the Tarmiya cell members drove north out of town in a convoy of pickup trucks, stopping only when they reached Al Qaim, the border town that was fast becoming the Islamic State's last redoubt in Iraq. They threw Harith into the Islamic State jail there, a fetid building where men were shackled with iron chains and left to rot. Sometimes they were fed, but some days not. Some prisoners wasted away from illness, some lost their minds, some died from the torture.

In June, the Falcons learned from another Islamic State detainee that Harith had suffered the same sort of treatment at the hands of a senior Islamic State member who had worked in Mosul with Abu Qaswarah. This man, Abu Thabit, was enraged over the way the war had turned against the extremists. Islamic State fighters were in a full retreat from all their major positions in Iraq; the coalition was decimating their forces with dozens of airstrikes a day, and the commander was incensed. Abu Thabit unleashed his temper against Harith.

The Falcons' detainee told Abu Ali that Harith endured unimaginable pain and prayed for death. But in the face of his tormentor, he barely talked, except to deny that he was a spy.

The Falcons never found a witness from the Al Qaim prison who could tell them how Harith made it through his final days. Two months later, in August, the Falcons received the final piece of evidence that erased all doubt about the fate of their man.

A colleague of Abu Ali's in the domestic intelligence agency sent the spy chief a video that his interrogators had recovered from another prisoner's phone. Entitled "Day of Retribution," the five-minute clip took place on what appeared to be a sandy bluff just outside of Al Qaim overlooking the plains to the west of the town past the thin, dirty ribbon of the Euphrates River that separates Iraq from Syria.

The intelligence officials could not tell if the media file had been created for widespread public consumption, as was normally the Islamic State's practice, or if it was a performance staged for a private audience, like the group's leadership. The production value was not as slick as most of the group's propaganda.

For those who have studied the videos of hundreds of executions released by the terror group, the one that the Falcons watched on August 18 was unremarkable. The soundtrack had a male voice reciting a Koranic prayer in a minor key, his voice digitally amplified to resemble a dirge. The group had arrayed eight prisoners dressed in orange jumpsuits with black balaclavas over their heads. Each man knelt on the sand, while unidentifiable jihadis, armed with pistols, stood behind them. Two of the captors also wore sword scabbards on their belts.

Another male voice cut into the prayer recital and announced a death sentence for the mutadeen—the traitors—who he declared had been officers of the Iraqi government. After that terse statement, the armed men executed the prisoners with shots to the head. The two extremists with swords then proceeded with the gruesome task of hacking the prisoners' heads from their bodies. The action takes an agonizingly long time. Blood pooling near the corpses starts to disappear as it seeps into the earth, when suddenly, and without warning, the feed goes dark.

Abu Ali and Munaf watched the video separately the first time, and then again together, once Munaf understood what he was seeing.

Munaf walked into Abu Ali's office struggling to control his tears. Almost every day for twenty years, he and Harith had slept in the same room in their father's house. He had grown up wearing his brother's hand-me-down clothes and sat next to him at almost every meal. The men being executed were not formally identified by their killers, but Munaf knew that the man on the screen, the second one on the left, was Harith. He had no doubt about it.

Abu Ali sat next to his young officer on the upholstered sofa in his office and watched him cry.

May God have mercy on him and help him enter paradise, the spy chief murmured to the grief-stricken brother. May God have mercy on us all.

Epilogue

The late-October evening breeze coming off the Tigris River carried the lingering warmth of a Baghdad summer. It was the fall of 2019, and lights twinkled across the Iraqi capital, wedding bands celebrated, and families rested peacefully in the still of the night, free of bombings or terror.

Abu Ali al-Basri stepped out of his stuffy office and breathed in the dusty scent of the city. The nightmare unleashed across Iraq when the Islamic State forcibly seized control of a third of the nation and threatened the lives of tens of millions was over. And he had kept his vow to the al-Sudani family. In the days after Harith had been kidnapped, Abu Ali had sworn to Munaf that he wouldn't rest until he found the men responsible for his brother's death.

In the early hours of October 27, under the darkness of a new moon, U.S. special forces killed Abu Bakr al-Baghdadi, the Iraqi who had led the most brutal terror group of the twenty-first century and who had destroyed so many lives.

It had taken almost two years of deliberate and painstaking intelligence-gathering to find the man ultimately responsible for all of the Islamic State's crimes, the Iraqi jihadi who had climbed from the ranks of an obscure religious scholar to his self-proclaimed but fallacious title of caliph, leader of the world's Muslims.

Throughout 2018, Abu Ali and his Falcons worked with the U.S. special forces team dedicated to tracking the onetime Islamic Studies teacher using the same painstaking incremental techniques that they had used a decade earlier to kill the top echelons of Al Qaeda in Iraq. Successful counterterrorism cases start with the small things, an accumulation of knowledge in the files that Abu Ali always keeps in hulking piles on his desk. In the spring and early summer, the Falcons' top interrogator who had worked with the Delta Force team involved in the deaths of Abu Omar al-Baghdadi and Abu Ayyub al-Masri flew with U.S. commandos between Iraq and Syria, scooping up detained Iraqi Islamic State fighters in search of more and more information, including which tribal network or family relationship could be leveraged for cooperation.

One of the first breaks in the hunt for the Islamic State leader came in February 2018. The Falcons told the Americans that they had discovered one of the leader's top aides living under an assumed name in the quiet conservative Turkish university town of Sakariya, a place where a Sunni Islamic scholar like himself could blend in.

Issam al-Ethawi was one of the oldest colleagues that al-Baghdadi had in the movement. He joined Al Qaeda in 2006 and was arrested by the American military in 2008. After the formation of the Islamic State, al-Ethawi lived in Mosul with the caliph and delivered high-level orders to his commanders. When the Iraqi security forces regained control over Mosul, al-Ethawi fled to Syria with his Syrian wife and then slipped across the border with his family into Turkey using his brother's identity papers.

Abu Ali gave the file on Issam al-Ethawi to the Americans, who propelled the case forward by pressuring the Turks to arrest the Iraqi and hand him back to the Iraqi authorities.

It took five months, but this is what the Iraqis coaxed out of him: the intelligence that al-Baghdadi communicated directly with only five men; the code words they used to mask their intentions; the routes these men used to smuggle themselves out of Mosul and across Syria to evade the Americans; and finally the location of safe houses that the men would be trying to use to keep out of harm's way.

Then the Falcons helped score an extra point. They convinced al-Ethawi to participate in a sting operation in May 2018, designed to lure four of the remaining inner circle, three other Iraqis and a Syrian, back to Iraq, where U.S. and Iraqi forces were waiting to arrest them.

From that point on, it was just a matter of when, not if, the allies would find al-Baghdadi. Through 2019, Abu Ali's top interrogator traveled back and forth to Syria with the group of Delta operatives as they inched toward their goal. In the middle of the year, intelligence led the Iraqis and Americans to conclude that their most-wanted man was hiding in Idlib, Syria, close to the Turkish border. The Americans took over the day-to-day surveillance of the area.

On October 26, when the countryside lay in darkness, eight American helicopters carrying a platoon of special forces took off from northern Iraq for the target area in Syria. In the village of Barisha, U.S. commandos, backed by air cover from the helicopters, stormed the austere single-story stucco villa where the Islamic State leader was hiding out with a dozen family members.

With the U.S. forces in hot pursuit, the Islamic State leader and two of his sons retreated into a network of underground bunkers and tunnels that snaked through the compound. When he hit a dead end, he detonated a suicide vest he had been wearing at the time of the attack, killing himself and his children.

Back in Baghdad, Abu Ali was gratified to hear of the undignified end to the brutish man.

It was a bittersweet moment for the spymaster—a moment to savor a significant victory in the protracted shadow war of counterintelligence and in his own quest to make Iraq a safer, more democratic nation. But also a moment of mourning and exhaustion, for Harith al-Sudani and for thousands of Iraqi families. The Islamic State killed an estimated 100,000 Iraqi civilians. Streets across the country were adorned with the faces of fallen soldiers. While the nation was infused with a new sense of patriotism, grief remained the dominant thread uniting Iraqis.

In Sadr City, in east Baghdad, that was certainly true for the al-Sudani family.

Since learning of his oldest son's death in 2017, Abu Harith had visibly withered. He had become sickly, and his usual erect posture had softened, as had his authoritarian manner, creating a jarring imbalance.

Instead of sitting in his armchair with a rigid back, Abu Harith took to reclining on a pallet in the al-Sudanis' majlis. The family patriarch still expounds on poetry and politics, but he has added a new topic to his conversations. Now, when neighbors and notables come to pay their respects, Abu Harith spends most of his time talking about the greatest spy in Iraqi history, his son.

Away from guests, however, Abu Harith becomes contemplative. Tears freely fall down his gaunt face as he counts up what he sees as his mistakes as a father. With the aching mix of worry and regret, he wonders whether Harith would still be alive if he had been a different kind of father, if he had shown him more affection, or done anything differently. By the time the family understood the nature of Harith's undercover work, his bravery and accomplishments, it was too late to tell his son he was proud of him, too late to say he loved him.

Upstairs in her own rooms, Raghad shares her father-in-law's remorse. The last conversation she had with her husband was bitter with recrimination. Had she known the importance of the mission that had kept him from his duties at home, from their children, she would have never yelled at him. She would have been more understanding, more supportive. Instead, he left the house for the final time, on his way back to a life behind enemy lines, angry and alone.

Munaf's guilt over allowing Harith to walk into the Islamic State's trap that cold January morning upended his life. In the two years it took to find and kill Abu Bakr al-Baghdadi, his wife gave birth to their first children, twin boys, and his professional accomplishments had earned him a promotion in rank. But from the moment he had viewed his brother's execution video, a blackness filled his heart. It was the same sensation Harith had described to him when he had told Munaf about the nightmare in which he was drowning. One month later, Munaf went to Abu Ali's office and tendered his resignation, telling the spymaster that he could no longer countenance the responsibility of holding agents' lives in his hands.

Abu Ali didn't have the psychological tools to deal with grief. In his world, men simply pushed through any emotional sludge. The only balm he could offer his young officer was his vow to get vengeance, and a recommendation for a new job as a detective with a new major crimes unit in Baghdad.

In west Baghdad, the al-Kubaisis, too, were a family unmoored in sorrow.

Professor al-Kubaisi spends most afternoons slumped in his beige upholstered armchair, his chin resting in his crooked left hand as weighty as a stone. The air in the living room is heavy. The sunlight pulsing outside in the garden barely reaches through the multiple layers of opaque gauzy curtains that hang over the windows like a shroud. He sits in this unfamiliar room in a strange neighborhood that he and his wife now call home, contemplating how the foundation of their lives had crumbled.

In the days following Abrar's arrest, the al-Kubaisis found themselves slowly isolated in their neighborhood, a silent social wall built upon the same fearful instincts bred during Saddam's republic of fear. None of the local families wanted their sons or daughters to be arrested for associating with the al-Kubaisis, and everyone assumed that the mukhabarat was still monitoring them. The family's next-door neighbor complained of the tens of thousands of dollars in damage to his home by the security forces during their overnight raid, and insisted that Professor al-Kubaisi pay for his repairs. For a family that held its reputation as dear as any material item, the stigma was unbearable.

Abrar's father put their home—the place that they had clung to like a life raft through the years of sectarian bloodletting and violence—up for sale. Now,

Baghdad was safer than it had been since 2003, but he and the family were leaving. His brothers found the family a new, much smaller home about seven miles away in a less prestigious neighborhood where no one would know of their shame.

Professor al-Kubaisi gazes blankly with his sightless eyes as he struggles to navigate unfamiliar emotions as well as the strange angles and hallways of his new home. For more than a year after her arrest, the al-Kubaisis refused to believe what Abu Ali al-Basri told them about their daughter—that she had joined the caliphate and plotted a terror attack against her hometown. That she wasn't the girl they thought they had raised. It was not until Abrar was moved out of her Baghdad holding cell to a secret prison in Erbil, in northern Iraq, where high-value terrorism detainees are interrogated by the Americans, that the young woman's parents realized the seriousness of her situation.

In the end, the family spent more than $10,000 on Abrar's legal defense, ignoring advice from several Iraqi attorneys that her case was hopeless. When the Americans are involved, they said, no one receives a get-out-of-jail-free card. The best that they could do was to plead with Baghdad's specialized terrorism court judge to spare her life. In September 2019, Abrar was convicted to life in jail. Um Mustafa visits her daughter once a month in Baghdad's maximum-security prison for women, where she is housed with more than thirty women in a cell. Some of them are like Abrar, she tells her mother, young pious women from good families. Some are illiterate villagers. The worst of the cell mates are hardened criminals, drug users, and murderers. Um Mustafa returns from these hour-long visits filled with despair. How could her clever girl have managed to ruin her life? How could she have thought that killing was something God wanted her to do?

Professor al-Kubaisi spends hours in his armchair trying to rationalize his reduced circumstance. But publicly he refuses to lay blame on his younger daughter. Instead, he finds comfort in the conspiracy theories that led to her radicalization all those years ago. Abrar, he says, was set up by the same Shiite oppressors who closed the doors on her professional dreams.

If Iraq could only return to its glory days, if only a strong man could rise up and take over and put those upstart Shiites in their rightful place, he muses, then maybe life could get better.

Back in the Falcons' headquarters, that late-October evening, Abu Ali sat alone in his office, his mind already turning toward the future. He didn't have patience for mourning or the distraction of grief. Instead of sadness, he prefers to concentrate on pride.

The spymaster of Baghdad gazed over his wire-rimmed glasses toward the overstuffed sofa, where four years earlier Harith had stood at attention announcing his desire to volunteer for his undercover mission. Abu Ali had four grown sons whom he loved dearly, but that show of courage by Harith al-Sudani had made him proud in a way that his own flesh and blood never had.

The threat of terrorism may have ebbed that week with the death of Abu Bakr al-Baghdadi, but Abu Ali could not afford complacency. The spy chief folded his hands on his desk and offered a silent prayer for the sacrifice of Captain al-Sudani and all the Iraqis who gave their lives in the battle against the Islamic State. Then he reached over and pulled open a new file, in search of the next mission to keep his homeland safe.

Acknowledgments

This book could not have been written without the unstinting help of many Iraqis, especially those who helped make Baghdad my adopted home. To my wonderful team—Ali Maliki, Ala'a, Ahmed, Bayan, Harith, Walid, and Walid Abu Mustaf—thank you for keeping me safe and in good cheer. To Khalid and Mohammed, for your diligence and dedication during our countless hours of interviews and months of research. To Hassan and Zahra'a for a remarkable road trip and much-needed dinner parties. To Lubna for your critical eye and inspiration.

I owe a special debt to the uncle of Baghdad's journalistic community, a man whose generosity, wisdom, and collegiality have no match. Thank you, Falih.

Thanks to the Georgia Arts Council and the Hambidge Center, the oldest artists' residency program in the Southeast. Foxfire, my cabin in the woods, was an appreciated rest stop on the way to this finish line.

I am grateful to my family, whose pride and support have sustained me during my long journeys to the unknown.

Finally: Anyck, Gonzague, and Margaret—through war zones and emotional trenches, you have been the best of friends. Thanks for the Paris cocoon and the oft-used London guest room, despite the ever-present risk of Swedish jazz interludes.

Notes on Interviews and Citations

Research for this book took place across Iraq, in London, and in Washington, D.C., between December 2018 and February 2020 and involved interviews with thirty-two current and former Iraqi and U.S. intelligence, military, and government officials and, separately, seven European and Mideast officials with direct knowledge of counterterrorism operations in Iraq. Some of the Iraqis interviewed include former prime ministers Haider al-Abadi, Nouri al-Maliki, and Ayad Allawi. Iraq's longtime national security adviser, Mowaffak al-Rubaie, the former State Minister for National Security Shirwan al-Waeli, and director of National Intelligence Mustafa al-Kadhimi were generous with their time and insight. Some of those interviewed agreed to be named in this book, while others, especially officials who remain in active service and involved in Iraqi intelligence operations, remain anonymous for security and operational reasons.

I owe a debt of gratitude to Abu Ali al-Basri and his colleagues in the Falcons Intelligence Unit who participated in more than ninety hours of interviews about their lives and work defending Iraq and its allies from terrorism.

I thank the al-Sudani family and the al-Kubaisi family for their trust and cooperation. Both extended to me the greatest hospitality and generosity during months of interviews in their homes. The al-Kubaisis provided me access to their daughter's legal documents as well as help communicating with Abrar in prison.

In addition to these primary sources, I also drew upon secret and classified reports of Iraqi intelligence and counterterrorism operations, interrogation records of Al Qaeda and Islamic State detainees, and minutes of interagency meetings between Iraqi, U.S., and other security officials. The official history of the Iraq War published by the U.S. Army War College in 2019, which drew on thirty thousand pages of declassified documents, was also invaluable.

The English translation of Qabbani's poem on p. 24 comes from Nizar Qabbani, Bassam Frangieh, and

Clementina R. Brown, *Arabian Love Poems: Full Arabic and English Texts* (London: Lynne Rienner Publishers, 1993).

The George H. W. Bush quote on p. 89 encouraging Iraqis to rise up against Saddam Hussein comes from remarks made by the president on February 16, 1991, to the American Academy for the Advancement of Science in Washington and to an audience at a Raytheon plant in Andover, Massachusetts.

The clandestine meeting during which Samarra was selected as the site of Al Qaeda's next terror attack is documented in Colonel Joel D. Rayburn and Colonel Frank K. Sobchak, *The U.S. Army in the Iraq War*, Vol. 1, pp. 531–532 (Carlisle, PA: U.S. Army College Press, 2019).

The quote from Abu Bakr al-Baghdadi's announcement of the Islamic State on p. 219 comes from Cameron Glenn, "The ISIS Primer," The Wilson Center, November 19, 2015, https://www.wilsoncenter.org/article/the-isis-primer.

General Ray Odierno's assessment of the Iraq war on p. 261 comes from his briefing on the state of the army in the Pentagon Press Briefing Room, August 12, 2015, Washington, D.C.

Bibliography

A'al, Ali Abdel. October 28, 2019. "Baghdadi, 2 wives, and guards killed in northern Syria raid." The Portal Center. https://www.theportal-center.com/2019/10/baghdadi-2-wives-and-guards-killed-in-northern-syria-raid/.

Abdel Zahra, Qassim. April 30, 2010. "Militant Led Iraqis to Al-Qaeda Chiefs." Associated Press. http://www.nbcnews.com/id/36876627/ns/world_news-mideast_n_africa/t/militant-led-iraqis-al-qaida-chiefs/#.XkGPUlJKjUr.

"Al-Qaeda: The Many Faces of an Islamist Extremist Threat." Report of the House Permanent Select Committee on Intelligence, 2006. https://fas.org/irp/congress/2006_rpt/hpsci0606.pdf.

Bakier, Abdul Hameed. May 2010. "Internet Jihadists React to the Deaths of Al-Qa`ida's Leaders in Iraq."

CTC Sentinel. Vol. 3, Issue 5. https://ctc.usma.edu/internet-jihadists-react-to-the-deaths-of-al-qaidas-leaders-in-iraq/.

Benraad, Myriam. June 2010. "Assessing AQI's Resilience After April's Leadership Decapitations." CTC Sentinel. Vol. 3, Issue 6. https://www.ctc.usma.edu/posts/assessing-aqis-resilience-after-aprils-leadership-decapitations/.

Bensahel, Nora, Olga Oliker, et al. 2008. "After Saddam: Prewar Planning and the Occupation of Iraq." (RAND Monograph MG-642-A.) Santa Monica, CA: RAND Corporation.

Bunzel, Cole. 2015. "From Paper State to Caliphate: The Ideology of the Islamic State." The Brookings Institution. Accessed June 15, 2017. https://www.brookings.edu/wpcontent/uploads/2016/06/The-ideology-of-the-Islamic-State.pdf.

Byman, Daniel. 2015. "Al-Qaeda, the Islamic State, and the Global Jihadist Movement: What Everyone Needs to Know." New York: Oxford University Press.

———. "The Resurgence of Al-Qaeda in Iraq." December 12, 2013. Testimony to a joint hearing of the Terrorism, Nonproliferation, and Trade Subcommittee and the Middle East and North Africa Subcommittee of the House Committee on Foreign Affairs. https://

www.brookings.edu/testimonies/the-resurgence-of-Al-Qaeda-in-iraq/.

Central Intelligence Agency. 2004. "Iraq's Intelligence Services. Regime Strategic Intent—Annex B." Special Adviser to the Director of Central Intelligence on Iraq's Weapons of Mass Destruction. https://www.cia.gov/library/reports/general-reports-1/iraq_wmd_2004/chap1_annxB.html.

Cohen, Zachery, and Barbara Starr. October 31, 2018. "Pentagon releases first images from raid that killed ISIS leader." CNN. https://edition.cnn.com/2019/10/30/politics/pentagon-baghdadi-raid-video/index.html.

Cordesman, Anthony, and Sam Khazai. July 3, 2012. "Iraq After US Withdrawal: US Policy and the Iraqi Search for Security and Stability." Center for Strategic and International Studies. https://csis-prod.s3.amazonaws.com/s3fs-public/legacy_files/files/publication/120702_Iraq_After_US_Withdrawal.pdf.

Cordesman, Anthony, and Adam Mausner. August 2009. "Withdrawal from Iraq: Assessing the Readiness of the Iraqi Security Forces." Center for Strategic and International Studies. https://csis-website-prod.s3.amazonaws.com/s3fs-public/legacy_files/files/publication/090812_Cordesman_WithdrawalIraq_Web.pdf.

Crocker, Ryan C. May 15, 2007. "Maliki Reshapes the National Security System." Department of State. https://wikileaks.org/plusd/cables/07BAGHDAD1593_a.html.

———. November 20, 2008. "Iraqi NSC Discusses NTM-I and Other Issues at November 16 Meeting." Department of State. https://wikileaks.org/plusd/cables/08BAGHDAD3671_a.html.

Department of Defense. June 2010. "Measuring Stability and Security in Iraq." Report to Congress. https://dod.defense.gov/Portals/1/Documents/pubs/June_9204_Sec_Def_signed_20_Aug_2010.pdf

Department of State. May 6, 2009. "Human Rights Vetting for US Sponsored Training for IP Integration NIIA—Group 3." https://wikileaks.org/plusd/cables/09BAGHDAD1194_a.html.

———. June 25, 2009. "Human Rights Vetting for U.S Sponsored Training for National Intelligence and Investigation Agency (NIIA)—Group 3." https://wikileaks.org/plusd/cables/09STATE66045_a.html.

Dobbins, James, et al. 2009. *Occupying Iraq: A History of the Coalition Provisional Authority.* [e-Book]. (RAND Monograph MG-847-CC.) Santa Monica, CA: RAND Corporation. http://www.rand.org/pubs/monographs/MG847.html.

Dulles, Allen W. 2016. *The Craft of Intelligence.* Guilford, CT: Rowman & Littlefield.

Elder, Gregory. "Intelligence in War: It Can Be Decisive." Center for the Study of Intelligence, Vol. 50, No. 2. Central Intelligence Agency. https://www.cia.gov/library/center-for-the-study-of-intelligence/csi-publications/csi-studies/studies/vol50no2/html_files/Intelligence_War_2.htm.

Fishman, Brian. August 2011. "Redefining the Islamic State: The Fall and Rise of Al-Qaeda in Iraq." National Security Studies Program Policy Paper. New America Foundation. https://static.newamerica.org/attachments/4343-redefining-the-islamic-state/Fishman_Al_Qaeda_In_Iraq.023ac20877a64488b2b791cd7e313955.pdf.

Frangieh, Bassam, and Clementina R. Brown, trans. 1999. *Arabian Love Poems,* by Nizar Qabbani. Full Arabic and English Texts. London: Lynne Rienner Publishers.

Gerges, Fawaz A. 2016. *ISIS: A History.* Princeton NJ: Princeton University Press.

Gold, Zach. October 2017. "Al-Qaeda in Iraq (AQI): An Al Qaeda Affiliate Case Study, Center for Stability and Development." Center for Strategic Studies. https://www.cna.org/CNA_files/PDF/DIM-2017-U-016118–2Rev.pdf.

Grey, Stephen. 2015. *The New Spymasters: Inside Espionage from the Cold War to Global Terror.* London: Penguin Books.

Hashim, Ahmed S. 2018. *The Caliphate at War: The Ideological, Organisational and Military Innovations of Islamic State.* London: Hurst.

Hassan, Hassan. (in press). *Tribes of ISIS: Infiltrating Terror Networks in the Middle East.* London: I. B. Tauris.

Izady, Michael. "Baghdad: Before and After the Sectarian Violence 2006–2007." The Gulf/2000 Project, School of International and Public Affairs, Columbia University, New York. https://brilliantmaps.com/baghdad-violence/.

Johnson, David, et al. 2013. "The 2008 Battle of Sadr City: Reimagining Urban Combat." (RAND Research Report RR-160-A.) Santa Monica, CA: RAND Corporation. https://www.rand.org/content/dam/rand/pubs/research_reports/RR100/RR160/RAND_RR160.pdf.

Khalil, Peter. July 2006. "Rebuilding and Reforming the Iraqi Security Sector." Saban Center for Middle East Policy, Brookings Institution. Analysis Paper Number 9. https://www.brookings.edu/wp-content/uploads/2016/06/khalil20060713.pdf.

Khalilzad, Zalmay. February 10, 2006. "Shia Islamist PM Contenders: No Perfect Candidate." Department of State. https://wikileaks.org/plusd/cables/06BAGHDAD402_a.html.

———. March 8, 2006. "Ambassador and PM: Neighbors Conference, Saudi Agenda, Kurdistan Trip." Department of State. https://wikileaks.org/plusd/cables/07BAGHDAD831_a.html.

———. September 27, 2006. "Iraqi Vice President Al-Hashemi Reviews Federalism, Security, Other Issues with Ambassador." Department of State. https://wikileaks.org/plusd/cables/06BAGHDAD3609_a.html.

———. October 10, 2006. "October 8 MCNS: Reform on the MIND." Department of State. https://wikileaks.org/plusd/cables/06BAGHDAD3757_a.html.

———. November 21, 2006. "Maliki Creates Crisis Cell; Invites Ambassador to Participate as Needed." Department of State. https://wikileaks.org/plusd/cables/06BAGHDAD4317_a.html.

———. November 28, 2006. "PM Discusses Intel Chief's Return to Iraq, Security Transfer with Ambassador." Department of State. https://wikileaks.org/plusd/cables/06BAGHDAD4387_a.html.

Knights, Michael. December 12, 2013. "The Resurgence of Al-Qaeda in Iraq." Report of the Joint Hearing of the Terrorism, Nonproliferation, and Trade Subcommittee and the Middle East and North Africa Subcommittee of the House Committee on Foreign Affairs. http://docs.house.gov/meetings/FA/FA18/20131212/101591/HHRG-113-FA18-Transcript-20131212.pdf.

Lamb, Christopher J., and Evan Munsing. March 2011. "Secret Weapon: High-value Target Teams as an Organizational Innovation." Center for Strategic Research, Institute for National Strategic Studies. Washington, DC: National Defense University Press.

Lamonthe, Dan, and Ellen Nakishima. October 27, 2019. "With Baghdadi in their sights, U.S. troops launched a 'dangerous and daring nighttime raid.'" *Washington Post.* https://www.washingtonpost.com/world/national-security/with-baghdadi-in-their-sights-us-troops-launch-a-dangerous-and-daring-nighttime-raid/2019/10/27/6c88a484-f8d0–11e9–9534-e0dbcc9f5683_story.html.

Lister, Charles. November 2014. "Profiling the Islamic State." Brookings Doha Center Analysis Paper, no. 13. Brookings Institution. https://www.brookings.edu/wpcontent/uploads/2014/12/ en_web_lister.pdf.

McCants, William. September 1, 2015. "The Believer." The Brookings Institution. http://csweb.brookings.edu/content/research/essays/2015/thebeliever.html.

McGurk, Brett. February 5, 2014. "Al-Qaeda's Resurgence in Iraq: A Threat to U.S. Interests." Report of the House Committee on Foreign Affairs, 113th Congress, Second Session. http://docs.house.gov/meetings/FA/FA00/20140205/101716/HHRG113-FA00-Wstate-McGurkB-20140205.pdf.

Milton, Daniel. July 2018. "Down, but Not Out: An Updated Examination of Islamic State's Visual Propaganda." Combating Terrorism Center at West Point, United States Military Academy. https://ctc.usma.edu/app/uploads/2018/07/Down-But-Not-Out.pdf.

Morell, Michael, and Bill Harlow. 2015. *The Great War of Our Time: The CIA's Fight Against Terrorism: From al Qa'ida to ISIS.* New York: Twelve.

Mosul Eye. June 2014. https://www.facebook.com/pages/Mosul-Eye/552514844870022?fref=nf.

Muir, Jim. April 26, 2009. "Top militant 'arrested in Iraq.'" BBC. http://news.bbc.co.uk/2/hi/middle_east/8019191.stm.

Nance, Malcolm. 2016. *Defeating ISIS: Who They Are, How They Fight, What They Believe.* New York: Skyhorse Publishing.

Orton, Kyle. May 10, 2018. "Rebel-Turned-Jihadist Saddam al-Jamal Reported Captured." https://kyleorton1991.wordpress.com/2018/05/10/rebel-turned-jihadist-saddam-al-jamal-reported-captured/.

Rasheed, Ahmed. October 27, 2019. "Baghdadi's aide was key to his capture—Iraqi intelligence sources." Reuters. https://www.reuters.com/article/us-mideast-crisis-baghdadi-capture-exclu/exclusive-baghdadis-aide-was-key-to-his-capture-iraqi-intelligence-sources-idUSKBN1X60SR.

Rathmell, Andrew, et al. 2005. "Developing Iraq's Security Sector: The Coalition Provisional Authority's Experience." RAND National Defense Research Institute, Arlington, Virginia. (Monograph MG-365.) https://www.rand.org/content/dam/rand/pubs/monographs/2005/RAND_MG365.pdf.

Rayburn, Col. Joel D., et al. January 17, 2019. "The U.S. Army in the Iraq War—Volume 2: Surge and Withdrawal, 2007–2011." Carlisle, PA: Army War College Press. https://publications.armywarcollege.edu/publication-detail.cfm?publicationID=3668.

Roggio, Bill. September 14, 2008. "Who is Abu Omar Al-Baghdadi." Foundation for Defense of Democracies (FDD), Long War Journal. https://www.longwarjournal.org/archives/2008/09/who_is_abu_omar_al_b.php.

Satterfield, David. July 5, 2006. "Ambassador Discusses Security, Services, and Kirkuk with Prime Minister Maliki." Department of State. https://wikileaks.org/plusd/cables/06BAGHDAD2354_a.html.

Scobey, Margaret. June 20, 2006. "Dawa Advisors Talk About PM Office, National Reconciliation Plan, Militias and COMSG." Department of State. https://wikileaks.org/plusd/cables/06BAGHDAD2099_a.html.

———. December 18, 2006. "Another Senior Dawa Figure Questions the Moderate Front." Department of State.

https://wikileaks.org/plusd/cables/06BAGHDAD4612_a.html

Special Inspector General for Iraq Reconstruction. March 2013. "Learning from Iraq." https://www.govinfo.gov/content/pkg/CHRG-113hhrg81868/pdf/CHRG-113hhrg81868.pdf.

Speckhard, Daniel, September 25, 2006. "September 24 Ministerial Committee on National Security." Department of State. https://wikileaks.org/plusd/cables/06BAGHDAD3587_a.html.

Swann, Glenn, et al. October 31, 2019. "Visual Guide to the raid that killed Isis leader Abu Bakr al-Baghdadi." *The Guardian.* https://www.theguardian.com/world/2019/oct/28/visual-guide-to-the-raid-that-killed-isis-leader-abu-bakr-al-baghdadi.

United Nations Mission for Iraq. "UN Casualty Figures for Iraq for the Month of October 2016." https://www.uniraq.org/index.php?option=com_k2&view=item&id=6267:un-casualty-figures-for-iraq-for-the-month-of-october-2016&Itemid=633&lang=en.

Warrick, Joby. 2015. *Black Flags: The Rise of ISIS.* New York: Doubleday.

———. 2011. *The Triple Agent: The Al Qaeda Mole Who Infiltrated the CIA.* New York: Anchor Books.

Weiss, Michael, and Hassan Hassan. 2016. *ISIS: Inside the Army of Terror*, rev. & updated. New York: Regan Arts.

Witty, David. March 16, 2015. "The Iraqi Counter Terrorism Service." Center for Middle East Policy, Brookings Institution. https://www.brookings.edu/research/the-iraqi-counterterrorism-service/.

About the Author

MARGARET COKER is a prize-winning investigative journalist who, for the last nineteen years, has covered stories from thirty-two countries on four continents. Since the U.S. invasion of Iraq in 2003, Coker has largely focused on the Middle East, writing on corruption, counterterrorism, and cyber warfare. Her stories written during the 2011 Libyan uprising over Muammar Gaddafi for the *Wall Street Journal* won prizes for investigative journalism and diplomatic reporting. As Turkey bureau chief for the *Wall Street Journal,* Coker contributed to a 2016 series that was a finalist for the Pulitzer Prize in International Reporting. Most recently, Coker was the *New York Times* bureau chief in Baghdad.